Ochre Pollution as an Ecological Problem in the Aquatic Environment
— Solution Attempts from Denmark

Die Edmund Siemers-Stiftung veröffentlichte bisher im Bereich Fließgewässerschutz

(Autor, soweit nicht anders angegeben: Dr. Ludwig Tent, Tostedt)

2006, Inga Krämer: Verrohrte Fließgewässer bei der Umsetzung der EG-Wasserrahmenrichtlinie – mögliche Lösungen und deren ökonomische Auswirkungen. BoD. – ISBN 10: 3-8334-6518-2, 13: 978-3-8334—6518-5.

2006: Ocker – ein Gewässerproblem, gegen das wir einiges tun können. – Ad fontes Verlag, Hamburg, 21 S., ISBN 3-932681-46-0

2002: Bessere Bäche – Praxistipps – Bereits geringer Aufwand bringt große Erfolge für den Lebensraum. – (gemeinsam herausgegeben mit: Hanseatische Natur- und Umweltinitiative Hamburg.) – Ad fontes Verlag, Hamburg, 68 S., ISBN 3-932681-3.

2001: Pflanzen und ihre Bedeutung für Fließgewässer – Praxistipps. – (gemeinsam herausgegeben mit: Hanseatische Natur- und Umweltinitiative Hamburg.) – Ad fontes Verlag, Hamburg, 52 S., ISBN 3-932 681-29-0.

2000, (Madsen, B. L. & L. Tent): Lebendige Bäche und Flüsse - Praxistipps zur Gewässerunterhaltung und Revitalisierung von Tieflandgewässern. Libri-BoD (Books on Demand), 156 S., ISBN 3-89811-546-1.

1998: Unsere Heidebäche brauchen Hilfe. Überarbeitete Neuauflage. Hamburg, 16 S. – ISBN 3-932681-30-4

Edmund Siemers-Stiftung, Schlankreye 67, 20144 Hamburg

Ochre Pollution as an Ecological Problem in the Aquatic Environment
– Solution Attempts from Denmark –

Verockerung als gewässerökologisches Problem
– Lösungsansätze aus Dänemark –

Hilke Prange

Dissertation work, october 2005 (Diplomarbeit)
University of Applied Sciences
Hochschule Bremen
Department 7 (Fachbereich 7)
International Degree Course Industrial and Environmental Biology
(Internationaler Studiengang für Technische und Angewandte Biologie, ISTAB)

Tutors (Referenten):
Prof Dr Heiko Brunken
Dr. G. Weber

Impressum

© 2007 Edmund Siemers-Stiftung, Hamburg

Herstellung und Verlag: Books on Demand GmbH, Norderstedt

ISBN
ISBN-10: 3-8334-6632-4
ISBN-13: 978-3-8334-6632-8

Umschlag-Layout: Holger Kurz und Ludwig Tent auf der Grundlage Madsen, B.L. & L. Tent (2000): Lebendige Bäche und Flüsse – Praxistipps zur Gewässerunterhaltung und Revitalisierung von Fließgewässern, ISBN 3-89811-546-1

Standardvermerk Der Deutschen Bibliothek:
Bibliographische Information der Deutschen Bibliothek:
Die Deutsche Bibliothek verzeichnet diese Publikation in der Deutschen Nationalbibliografie; detaillierte bibiliografische Daten sind im Internet über <http://dnb.ddb.de> abrufbar.

This dissertation work was supported by (diese Diplomarbeit wurde unterstützt von):

Edmund Siemers-Stiftung
Schlankreye 67
20144 Hamburg
Germany

Sønderjyllands Amt
Teknisk forvaltning, Vandløbsafdelingen
Jomfrustien 2
6270 Tønder, Denmark

Hilke Prange

Permanent address:
Blenhorsterstr.38
31613 Wietzen

E-mail: hilkeprange@gmx.de
Mobile: 0179 / 58 74 92 8

Table of Contents

1. Introduction 1
1.1 Background and motivation of this dissertation 1
1.2 Investigation areas 2
1.3 Landscape development 4
1.4 Influence of ochre pollution on biological quality parameters of the WFD 4
1.5 Natural abundance of iron within terrestrial and aquatic environments 6
1.6 Chemical iron oxidation 7
    1.6.1 A redox-reaction determines the relation between ferrous and ferric iron 7
    1.6.2 Chemical iron oxidation and its interaction with chemical parameters 8
1.7 Biochemical iron oxidation 10
    1.7.1 Iron oxidation at springs 10
    1.7.2 Oxidation of pyrite within the soil 10
    1.7.3 Sediments and resuspension within standing waters 12
1.8 Different reasons of ochre pollution 13
    1.8.1 Anthropogenic reasons for ochre pollution 13
    1.8.2 Allochthone and autochthone ochre pollution 14
1.9 Aim of investigations 15

2. Materials and methods 16
2.1 General approach 16
2.2 Detailed investigations of three ochre lakes in the SJA 16
    2.2.1 Structure of the investigation of ochre lakes 16
    2.2.2 Collection of additional background information 17
    2.2.3 Investigation of iron concentrations and influence of discharge 17
    2.2.4 Investigation of iron loads 18
    2.2.5 Investigation of abundance and distribution of vegetation and turbidity 18
    2.2.6 Investigation of deposition, retention time and surface-volume-ratio 18
    2.2.7 Determine discharge within the vegetation channel at Hvirlå Ochre Lake 20
    2.2.8 Data investigation of oxygen, pH, alkalinity and temperature 20
    2.2.9 Equipment for the investigations within the field 21

3. Results 22
3.1 Abatement measures and strategies from Denmark 22
    3.1.1 Identification of ochre potential areas and threshold values 22
    3.1.2 Approval of drainage projects and stream targets 23
    3.1.3 Where to start – the priority list of the Sønderjyllands Amt 25
    3.1.4 Review of ochre projects 26
3.2 Parameters that influence instalment and maintenance of ochre lakes 26
    3.2.1 Definition of the term "ochre lake" 26
    3.2.2 Volume and retention time 26

    3.2.3 Position within the stream system 27
    3.2.4 Maintenance and disposal 27
    3.2.5 Vegetation within the ochre lakes 28
    3.2.6 Attempts to improve cleaning efficiency 29
  3.3 Examples for different types of lakes 30
  3.4 Definition of efficiency 32
  3.5 Detailed investigations of three ochre lakes in the SJA 33
    3.5.1 Hvirlå Ochre Lake 33
    3.5.2 Løgumkloster Bæk Ochre Lake (dam across a valley) 42
    3.5.3 Landeby Bæk Ochre Lake 49
  3.6 Further possibilities to combat ochre pollution 57
    3.6.1 Winter Ochre Lakes 57
    3.6.2 Combat the origin of ochre pollution – raise water tables 58
    3.6.3 Ochre abatement projects within clearly demarked valleys 58
    3.6.4 Ochre abatement measures within flat terrain 61
    3.6.5 Alteration of maintenance 61

4. Discussion 67
  4.1 Know how and experience are already available 67
  4.2 Investigation of the cleaning efficiency of three ochre lakes in the SJA 67
    4.2.1 Impact of the catchments area on amounts of iron that reach the ochre lake 67
    4.2.2 Reasons for the different cleaning efficiencies within the investigated ochre lakes 68
    4.2.3 Consequence for the aquatic ecosystem downstream the ochre lakes 72
  4.3 Combat reasons rather than symptoms – an outlook 74
    4.3.1 Ochre lakes in comparison to other measures 74
    4.3.2 Application of already abundant know how? 75

5. Literature 77

Vorwort

Gewässerschutz für Bäche und kleine Flüsse

Die Edmund Siemers-Stiftung engagiert sich seit ihrer Gründung im Naturschutzjahr 1995, das im Zeichen von „Naturschutz außerhalb von Schutzgebieten" stand, unter anderem konsequent für Verbesserungen an Bächen und kleinen Flüssen. In einem eigenen, langfristig angelegten Naturschutzprojekt, das die Vernetzung der Gewässerlandschaften im Oberlaufbereich der Heidebäche Seeve, Este, Oste und Wümme zum Ziel hat, werden best practice-Beispiele umgesetzt, die Fachleuten und Zuständigen als Anschauungsobjekte dienen können. Die Bedeutung gerade der Bäche und kleinen Flüsse, die mehr als 80 % der Fließstrecken repräsentieren, das Erkennen der gegenwärtigen lebensraumfeindlichen Situation und die notwendigen Veränderungen täglichen Handelns werden verdeutlicht – immer begleitet vom Hinweis auf die erforderliche Wahl des richtigen, standorttypischen Leitbilds. Im Rahmen der zwischenzeitig in Kraft getretenen Wasserrahmenrichtlinie haben bereits eine Vielzahl Besucher aus Gewässerunterhaltung, Wasser- und Naturschutzverwaltungen, Bildungseinrichtungen und der interessierten Öffentlichkeit dieses Angebot genutzt. Sowohl die Fließgewässer des Norddeutschen Tieflands als auch Flachlandgewässer anderer Regionen Europas profitieren inzwischen von diesen Erfahrungen.

Daneben ist es Anliegen der Stiftung, Beispiele guter fachlicher Praxis durch Vorträge, Exkursionen und Schriften zu vermitteln. Dies geschieht unter anderem durch Übersetzungen allgemeinverständlicher Texte aus dem Dänischen – hat doch Dänemark seit mehr als 25 Jahren einen Schwerpunkt auf lebensraumbezogene Gewässerverbesserung gelegt und steht beispielhaft im europäischen Vergleich.

Ein wichtiges Anliegen der Edmund Siemers-Stiftung ist weiterhin die Bildungsarbeit für nachhaltige Entwicklung im Gewässerschutz. So werden umsetzungsorientierte Themen an Fachhochschulen und Hochschulen gefördert mit dem Ziel, notwendige Handlungsoptionen für Fachwelt und Öffentlichkeit besser zugänglich zu machen. Die neue Schriftenreihe ergänzt dieses Vorgehen, hierin werden in lockerer Folge aktuelle Themen dargestellt. Besonderer Fokus wird auf Themen gelegt, die andernorts (noch) nicht hinreichend berücksichtigt sind.

Die vorliegende Arbeit von Hilke Prange stellt ein in Deutschland vielerorts vernachlässigtes Thema des Gewässerschutzes dar: die vielfältigen, unter anderem bei Störungen des Bodenwasserhaushalts auftretenden Probleme der Verockerung und Möglichkeiten zu ihrer Lösung. Durch unterschiedliche Ursachen, unter anderem durch Dränung von Flächen freigesetztes gelöstes Eisen wird den Gewässern zugeführt, dort in unlösliche Verbindungen oxidiert und schlägt sich dort nieder. Hierdurch wird z.B. der ökologisch so wichtige Porenlückenraum der Laich- und Aufwuchsbiotope vieler Gewässerorganismen gravierend beeinträchtigt, auch die Wasserpflanzen werden in Arten- und Individuenzahl erheblich gestört. Diese chronische Belastung, oft genug durch Dränspülungen akut verstärkt, verhindert also dauerhaft ein vielfältiges Leben im Bach und das Erreichen der standorttypischen Produktivität.

Neben dem Eingriff in den Bodenwasserhaushalt ist unter anderem die nicht angepasste, harte Gewässerunterhaltung eine weitere, wesentliche Ursache für eine Problemverschärfung. Extreme Spitzen an Eisenocker-Konzentrationen belasten Tier- und Pflanzenwelt, Strukturvielfalt, Produktivität und Selbstreinigungsleistung der Gewässer werden in ihrer Entwicklung behindert.

Dänemark hat seit Jahren, gestützt durch ein spezielles Ockergesetz, systematisch an dieser Thematik gearbeitet. Möglichkeiten, das Problem zu minimieren, wurden gesucht und gefunden. In Zusammenarbeit mit dänischen Ämtern und Fachleuten hat Hilke Prange eine Arbeit vorgelegt, die es in dieser fachübergreifenden Darstellung und Gesamtsicht in Deutschland bisher nicht gibt. Sie ist in englischer Sprache mit deutschen zusammenfassenden Informationen angelegt, da es sich um ein bedeutendes internationales Thema handelt.

Diese fast monographieartige Gesamtschau von Hilke Prange schließt eine Lücke, mit hohem Nutzwert insbesondere auch für den Praktiker vor Ort. Im Hinblick auf die Anforderungen der Wasserrahmenrichtlinie liegt ein wichtiger Beitrag für den Weg zum guten Zustand bzw. guten ökologischen Potential hiermit vor.

Die Edmund Siemers-Stiftung freut sich, diese, ein bisher zu wenig beachtetes Thema behandelnde, praxisorientierte Arbeit zu veröffentlichen und wünscht den Gewässerlebensräumen eine zügige und flächendeckende Umsetzung der dargestellten Verbesserungsmöglichkeiten.

Die aus Platz- und Übersichtsgründen hier nicht enthaltenen Anlagen der Diplomarbeit (Karten, Datengrundlagen) können bei Interesse als CD bei der Edmund Siemers-Stiftung bestellt werden.

Dr. Ludwig Tent   Dr. Andreas Wass von Czege

Abstract

This dissertation work was written during the 8th semester of the International Degree Course of Environmental Biology at the University of Applied Sciences Bremen (Hochschule Bremen). It is the continuation of the project work during the 7th semester in which the ecological importance of the ochre pollution for the implementation of the EU Water Framework Directive (WFD) was investigated and proved. The introduction of this dissertation work reviews and deepens results about the chemical and biochemical processes and anthropogenic reasons for ochre pollution. Ochre pollution occurs because of anthropogenic altered hydrological processes. Iron compounds and especially pyrite (FeS_2) are exposed suddenly to aerobic conditions. Chemical or biochemical processes alter former stabile compounds to mobile ferrous iron. These processes are based on the redox-reaction of iron, which is influenced by local physico-chemical parameters and therefore landscape development and land use. Ochre pollution starts in the catchments area. The biological quality parameters of the WFD (fish, benthic invertebrates, macrophytes, phytobenthos and phytoplankton) are influenced negatively by ochre pollution in several ways.

The occurrence of ochre pollution within watercourses of N/W-Germany has not been recognized as an ecological problem in aquatic ecosystems, yet. This dissertation work emphasized to collect and translate already abundant know how from Jutland/Denmark where ochre pollution has been combated over more than twenty years now. While the Ochre Act from 1985 aimed the legislation of new drainage projects, recent ochre pollution is mostly diffuse and caused by intensive maintenance. It occurs permanently.

The instalment of ochre sedimentation basins (so called ochre lakes) within ochre polluted streams is one possibility to diminish ochre pollution. During the ten weeks stay at the Sønderjyllandsamt (SJA) in South Jutland/Denmark three ochre lakes were investigated in detail to determine main influences that cause different cleaning efficiencies. The investigation was subdivided into three parts. The first part was to review reasons for ochre pollution in the catchments area. The cleaning efficiency within the ochre lake was determined by theoretical and practical investigations of iron concentrations, iron loads, reduced volumes and retention times, vegetation cover, channel development and alteration of physico-chemical parameters. Finally the positive and negative influences of the ochre lakes on the streams ecosystem were considered.

During excursions to the Ringkøbing County, Holstebro Municipality and Herning Municipality additional information could be gained about the installation of different types of ochre lakes and its maintenance. Shorter retention times because of reduced volume or channel development within the vegetation are the main reasons for reduced cleaning efficiencies here. Nevertheless vegetation increases oxygen and pH and therefore iron oxidation. Oxygen concentrations and pH decreased over years upstream two of the investigated ochre lakes. Oxygen concentrations are increased by these two ochre lakes, whereas high contents of organic matter can cause decreased oxygen supply at Løgumkloster Bæk Ochre Lake. The natural high alkalinity prevented also extreme alterations of the pH within the ochre lakes. Increased temperatures of approximately 4°C occur occasionally during summer season. Ferrous iron concentrations at two ochre lakes still exceed threshold limits but ochre lakes and other measures led to an improved water quality and made biological determination of water quality within former ochre polluted watercourses possible.

This dissertation work includes also descriptions of the identification of ochre potential areas, the determination of pyrite threshold limits, the involvement of threshold limits for ferrous iron into stream targets and the development of a priority list of ochre polluted streams within the Sønderjyllands Amt.

Finally other projects are described, e.g. the Ravsted Winter Ochre Lake and valley projects which aim(ed) to raise the groundwater tables again and combat the reason of ochre pollution in the catchments area. The enhancement of ochre pollution by intensive maintenance has been underlined by investigation results from Holstebro Municipality. Manual weed cut and favouring typical plant species is also described as a fundamental measure. Ochre pollution caused by intensive maintainance will be diminished, while an enhanced structural diversity and an improved self-cleaning efficiency are elemental aspects to reach good water status as demanded by the WFD.

Zusammenfassung

Diese Diplomarbeit wurde im 8. Semester des Internationalen Studienganges Umweltbiologie and der Hochschule Bremen geschrieben. Es ist die Fortsetzung der Projektarbeit im 7. Semester, in welcher die ökologische Bedeutung der Verockerung und ihre Relevanz für die Umsetzung der EG-Wasserrahmenrichtlinie untersucht und bewiesen wurde. Die Einleitung dieser Diplomarbeit zeigt nochmals die an der Verockerung beteiligten chemischen und biochemischen Prozesse auf. Verockerung als gewässerökologisches Problem entsteht durch die vom Menschen verursachte Veränderung hydrologischer Prozesse. Eisenver-bindungen und insbesondere Pyrit (FeS_2) werden plötzlich aeroben Bedingungen ausgesetzt. Chemische oder biochemische Prozesse verändern zuvor stabile Verbindungen in gelöstes zweiwertiges Eisen. Diese Prozesse basieren auf einer Redox-Reaktion, welche durch die lokalen physikalisch-chemischen Parameter und damit durch die Landschaftsentstehung sowie Landnutzung beeinflusst werden. Das Problem der Verockerung beginnt im Einzugsgebiet. Die biologischen Qualitätskomponenten der EG-Wasserrahmenrichtlinie werden vielfach negativ beeinflusst.

Die Erscheinung der Verockerung in Gewässern N/W-Deutschlands ist scheinbar bisher nicht als gewässerökologisches Problem erkannt worden. Der Schwerpunkt dieser Diplomarbeit liegt auf dem Zusammentragen bereits vorhandenen Wissens in Jütland/Dänemark, wo Forschung und Maßnahmen zur Minderung der Verockerung bereits seit zwanzig Jahren existieren und weiter entwickelt werden. Während das Ocker Gesetz die Administration aktueller Drainagevorhaben regelt, ist die heute auftretende Verockerung zumeist diffus und durch intensive Gewässerunterhaltung verursacht. Sie tritt permanent auf.

Die Anlage von Ockersedimentationsbecken (so genannten Ockerseen) im Verlauf der verockerten Gewässer ist _eine_ Bekämpfungsmöglichkeit. Während des zehnwöchigen Aufenthaltes im Sønderjyllands Amt in Süd Jütland/Dänemark, wurde die Reinigungseffizienz von drei verschiedenen Ockerseen untersucht. Die Untersuchung bestand aus drei Abschnitten. Zu Beginn erfolgte die Recherche zu den Ursachen der Verockerung im Einzugsgebiet des Ockersees. Die Reinigungseffizienz der Ockerseen wurde sowohl durch theoretische als auch praktische Untersuchungen der Eisenkonzentrationen, der Eisenfrachten, des reduzierten Volumens und der Retentionszeiten, sowie veränderte physikalisch-chemische Parameter untersucht. Anschließend wurden die positiven und negativen Einflüsse der Ockerseen auf das aquatische Ökosystem erwogen.

Während zusätzlicher Exkursionen in das Ringkøbing Amt, die Holstebro Kommune und die Herning Kommune konnten weitere Informationen über die Anlage und Unterhaltung von Ockerseen gewonnen werden. Verkürzte Retentionszeiten auf Grund verringerten Volumens und Stromrinnenbildung sind hier die Hauptursachen für die verringerte Reinigungsleistung. Nichts desto trotz erhöht die Vegetation den Sauerstoffgehalt und auch den pH-Wert und treibt damit die Oxidation des zweiwertigen Eisens voran. Sauerstoffkonzentrationen und pH sanken über Jahre hinweg stromaufwärts zwei der untersuchten Ockerseen. Diese beiden Seen hoben die Sauerstoffgehalte an, während im Løgumkloster Bæk Ockersee hohe Gehalte an organischer Substanz den Sauerstoff gelegentlich stark zehren können. Die natürlich hohe Alkalinität verhinderte extreme Veränderungen des pH-Wertes in den Seen. Temperaturerhöhungen um ca. 4°C traten vereinzelt in den Sommerperioden auf. Die Grenzwerte für gelöstes zweiwertiges Eisen werden bei zwei der Ockerseen überschritten, jedoch haben sie und andere Maßnahmen zu einer Verringerung der Verockerung geführt und die biologische Gewässergütebestimmung in zuvor stark verockerten Gewässern überhaupt erst möglich gemacht.

Diese Diplomarbeit beinhaltet außerdem die Beschreibung über die Identifikation verockerungsgefährdeter Gebiete, das Feststellen von Pyrit-Grenzwerten, das Einbeziehen von

Grenzwerten für gelöstes zweiwertiges Eisen in die Gewässerzielsetzungen und das Erstellen der Prioritätenliste verockerter Gewässer im Sønderjyllands Amt.

Zu guter letzt werden auch andere Möglichkeiten beschrieben, wie der Ravsted Winterockersee und verschiedene Autalprojekte die es u.a. zum Ziel haben (bzw. hatten) den Grundwasserflurabstand anzuheben und damit die Ursache der Verockerung zu bekämpfen. Die Verstärkung der Verockerung in intensiv unterhaltenen Gewässern wurde mit Untersuchungsergebnissen aus der Holstebro Kommune belegt. Die Gewässerunterhaltung von Hand und das Fördern gewässertypischer Pflanzenarten werden als eine weitere entscheidende Maßnahme beschrieben. Während die durch die intensive Gewässerunterhaltung verursachte Verockerung vermindert wird, sind Strukturvielfalt und eine verbesserte Selbstreinigungskraft entscheidende Bausteine für das Erreichen des guten ökologischen Zustandes.

1. Introduction

1.1 Background and motivation of this dissertation

This dissertation work was written during the 8th semester of the International Degree Course "Environmental Biology" at the University of Applied Sciences Bremen (Hochschule Bremen). Already during the 7th semester of this degree course a project report was written which reviewed "The ecological importance of ochre pollution and its relevance for the implementation of the EU Water Framework Directive" (Die ökologische Bedeutung der Verockerung und ihre Relevanz für die Umsetzung der EG-Wasserrahmenrichtlinie). Chemical and biochemical processes that cause ochre pollution and reasons for its appearance in N/W-Germany were investigated during literature research, several interviews and practical investigations at the Immer Bäke, a small watercourse in south-western direction of Bremen.

The negative impact of ochre pollution on the biological quality parameters, especially fish and benthic invertebrates, was proved and underlined the relevance of ochre pollution to the implementation of the Water Framework Directive (WFD). This directive demands to achieve and maintain the "good water status" of surface waters. The classification of water bodies is based on initial characterization and typology. This means the state in which organisms and substances would occur without any anthropogenic influences (UBA 2004). Drainage to reclaim arable land was identified as one main reason of ochre pollution. During the project work it has been also found out that ochre pollution has been tackled as an obvious ecological problem of watercourses in Denmark since the beginning of the 1980's.

The Sønderjyllands Amt (SJA) was contacted first in spring 2005. Mr Ludwig Tent suggested this dissertation work to the Edmund Siemers-Foundation which offered financial support. The core of this work was an investigation phase with duration of ten weeks (from 18th April until 30th June 2005) at the Sønderjyllands Amt (SJA) in South Jutland, Denmark. This dissertation work had two main goals. Already abundant experiences of ochre abatement measures in Denmark should be collected, summarized and translated into English (and German). Additional interviews and site visits at Ringkøbing County, Holstebro Municipality and Herning Municipality in the North of Jutland made it possible to gain more knowledge about different origins of ochre pollution and possible combat measures.

The instalment of sedimentation lakes within ochre polluted streams is one possible measure. The investigation of the cleaning efficiencies at different lakes at the SJA was the second part of this dissertation work. It describes origins of the aquatic ochre problem in detail and provides concrete solution attempts in an international context. Its aim is to initiate at least the consideration of ochre pollution as an ecological problem within watercourses of N/W-Germany and offers the eventual inclusion of already abundant experiences into the programmes of monitoring and measures as demanded by the WFD until 2009, 2012 respectively.

- Der negative Einfluss der Verockerung auf die biologischen Qualitätsparameter, sowie ihre Relevanz für die Umsetzung der EG-Wasserrahmenrichtlinie wurden in der Projektarbeit des 7. Semesters Umweltbiologie erarbeitet.
- Da die Verockerung bereits seit Anfang der 1980er Jahre in Jytland/Dänemark bekämpft wird, war der Kern dieser Diplomarbeit ein zehnwöchiger Aufenthalt in der Fließgewässerabteilung des Sønderjyllands Amtes in Jytland, Dänemark.
- Herr Dr. Ludwig Tent stellte den Kontakt zur Edmund Siemers-Stiftung her, welche diese Diplomarbeit finanziell unterstützte.

- Diese Diplomarbeit hatte zum einen das Ziel, bereits bestehendes Wissen über die Bekämpfung der Verockerung zusammenzutragen. Zum anderen sollte die Reinigungseffizienz verschiedener Ockerseen untersucht werden.
- Damit liefert diese Diplomarbeit Lösungsansätze aus der dänischen Praxis für das gewässerökologische Problem der Verockerung, welches in die Umsetzung der EG-Wasserrahmenrichtlinie einbezogen werden muss.

1.2 Investigation areas

Denmark is subdivided into 14 countys and 275 municipalities. Investigations of ochre lakes took place in the South Jutland County (Sønderjyllands Amt, SJA). It is the most southern part of Jutland and Denmark (figure 1). The SJA includes 23 municipalities. The Rivers **Vide**, **Brede** and the upper catchment area of the **Ribe** are the three main catchments areas that are abundant here. The total length of watercourses is approximately 14.000 km.

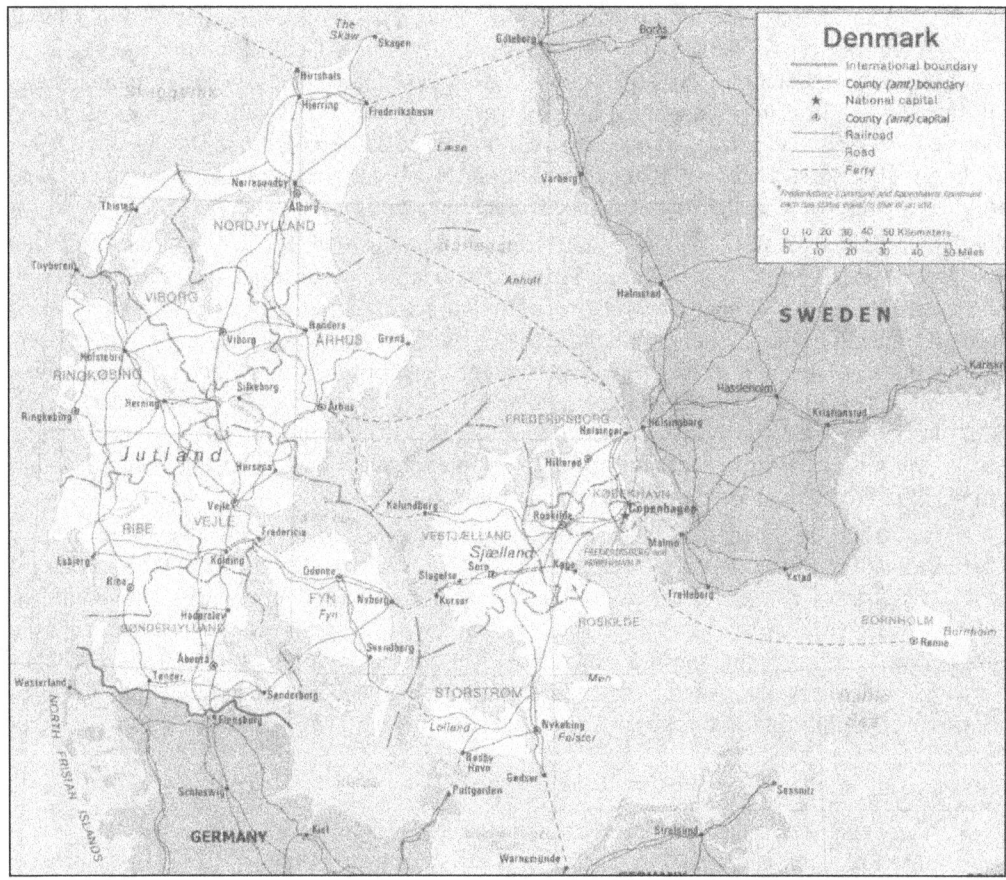

Figure 1[1]: South Jutland County (SJA) is located directly at the Danish-German border. Ringkøbing County, including Holstebro and Herning Municipalities is located in N/W-Jutland.

---

[1] URL: http://www.weltkarte.com/europa/landkarte_daenemark.htm [11.08.2006]

The **Vide** flows partly on the Danish-German border. Its catchments size is 14.000 km². The Hvirlå is part of this stream system. Regulation and drainage in the upper part of the Hvirlå during the 1950-75´s caused lower groundwater tables and leaching of ferrous iron. The biggest drainage projects took place at wet meadows and wetlands between Nørre Ønlev and Kasso (WANDALL & WHITELAW CHRISTENSEN 1999).

The **Brede** drains a catchments area of 473 km² and encompasses over 1000 km open canals, ditches and streams that enter the North Sea at Ballum flood gate. Several wetlands occurred here naturally (e.g. Kongens Mose and Alslev Mose). From the 1950´s to the 1970´s channels were straightened and wetlands were drained to reclaim arable land. Within areas with a high content of turf this led to decay of organic material and following lowered terrain (Moorsackung). From 1990 to 1998 several restoration projects aimed to improve structural as well as chemical water quality of the Brede and its northern extension, the Lobaek. A stretch of 19 km was prolonged to 25 km by remeandering. Different projects aimed to remove obstacles for free passage, better hydrological interaction between river and adjacent areas (flooded meadows), reduction of sand load and a decreased ochre pollution (GRØN 2000).

The **Ribe** is the third main river system with a catchments size of 1100 km². The SJA is responsible for the upper part of this waterway system. Site visits and interviews took also place in the Ringkøbing County, involving Holstebro and Herning Municipalities. These are located in the north-western part of Jutland (figure 1).

- Die Untersuchungen der Ockerseen fanden im Sønderjyllandsamt statt, in welchem die drei größten Einzugsgebiete die der Flüsse Vide, Brede und Ribe sind.
- Gewässerregulierung und Entwässerung von Feuchtgebieten fanden mit dem Ziel der Landgewinnung in den 1950 - 1970er Jahren statt.
- Das Restaurierungsprojekt des Flusses Brede beinhaltete u.a. die Bekämpfung der Verockerung.
- Weitere Informationen konnten während Exkursionen zum Ringkøbing Amt und den Kommunen Herning und Holstebro im Nord-Westen Jütlands gewonnen werden.

1.3 Landscape development

In general geology and pedology are the natural origin for chemical characteristics of surface waters and its catchments. The peninsula of Jutland was mainly formed during the last two ice ages. During the Saale ice age 100.000 years ago Jutland was completely covered by glaciers that transported a huge range of different geological materials from the North of Scandinavia. These formed a typical classification of natural landscapes that is described as "glacial series".

Glaciers of the last ice age (Weichsel ice age, 50.000 – 15.000 years ago) covered only the east of the peninsula and left a mountain range running from north to south (end moraine). Melting waters deposited lots of **sand** on the central part of South Jutland (Sander). Originally soils were poor of lime and nutrients here. Slowly melting glaciers in the east left moraine materials rich in lime, clay and loam (Geschiebelehm and Mergel). This caused fertile soils with higher amounts of lime ($CaCO_3$) and a natural high alkalinity (MILJØSTYRELSEN et al. 2004).

In comparison to that the **old moraines** in the west were not covered by protecting ice caps. Precipitation weathered rocks and leached out lime. Clay particles were blown away (SCHMITKE 1985, POTT 1999, KÜSTER 1999). Because of the low **alkalinity** and low **pH** streams are polluted in another way by ochre than within areas with high alkalinity and pH. Landscape development is therefore determining were and how intensive ochre pollution occurs depending on the origin of iron compounds as well as chemical parameters within soil and water. The situation in South Jutland and Ringkøbing County needs to be distinguished. Although land use and application of fertilizers has altered the landscape and chemical characteristics[2], the natural history provides necessary background knowledge.

- Die landschaftliche Entstehung beeinflusst Intensität und Phänotyp der Verockerung.
- In den Jungmoränen im Osten Jütlands tritt die Verockerung auf Grund höherer Kalkgehalte und höherer Alkalinität eher selten auf (natürlicherweise an Quellen).
- Damit ergeben sich unterschiedliche Situationen in den westlichen bis nord-westlichen Gebieten und den östlichen Gebieten Jütlands.

1.4 Influence of ochre pollution on biological quality parameters of the WFD

The WFD aims to achieve and maintain "good ecological status" for surface waters and "good quantitative status" for groundwater. The good ecological status is defined by the biological quality parameters. The good status as it is described for the biological parameters has highest priority. Hydromorphological (e.g. hydrological regime) and physico-chemical elements (e.g. oxygen balance and acid neutralizing capacity) are defined indirectly which means that they have to support the biological elements (Appendix V, Number 1.1 of the Directive 2000/60/EG). The importance of ochre pollution for the implementation of the WFD has been investigated in detail in PRANGE (2005). **Biological parameters** are "composition and abundance of aquatic flora" as well as "benthic invertebrate fauna" and "composition, abundance and age structure of fish fauna within rivers". Within lakes the "composition, abundance and biomass of phytoplankton" has to be monitored additionally (Appendix V, Number 1.1 of the Directive 2000/60/EG). Following paragraphs will review and deepen the information of the project report.

---

[2]URL: http:// www.umwelt.schleswig-holstein.de/servlet/is/24417/fglandschaften [15.07.05]

Aquatic flora includes macrophytes and phytobenthos. Both macrophytes and phytobenthos are stressed by low pH, high concentrations of sulphate and of heavy metals. The higher turbidity prevents photosynthesis and growth of macrophytes and algae. HASLAM (1987) reported about dominating *Sparganium emersum* and *Glyceria fluitans*. Phytobenthos is also covered by ochre which decreases abundance of benthic organisms that graze on it. In the early 1980's the average concentration of ferrous iron was 2.5 mg/l within many streams of South Jutland. This made evaluation of biological water quality (biologische Gewässergütebestimmung) impossible.

Numbers of **benthic invertebrates** that feed on algae are even reduced by ferrous iron concentration of more than 0.3 mg/l (REKER 1984). *Plecoptera* spec. and *Ephemeroptera* spec. are most sensitive to ferrous iron concentrations < 0.25 mg/l (SJA a 2005). Ochre polluted stretches are dominated by species whose reproduction takes place within two or more generations during summer season (REKER 1984). Ochre particles clog the interstitial and reduce therefore living spaces for benthic organisms.

If ferrous iron is precipitated on the gills **fish** will die of **suffocation** ("Ockerstrangulation"). The lower the pH the more toxic is the ferrous iron. At pH 6.5 – 7.0 ferrous iron should not be higher than 1.5 mg/l to maintain the population. At pH 6.0 it should not be higher than 0.5 mg/l (REKER 1984). If ochre is deposited on fish eggs respiratory cells are prevented to exchange gases. Eggs and larvae of trout (*Salmo trutta*) can not survive at concentration > 0.5 mg/l ferrous iron (JENSEN et al. 2003). Larvae that hatch under ochrous conditions suffer malformations.[3] BAUR (1997) even recommends less than 0.1 mg/l ferrous iron for fry and 0.3 - 0.5 mg/l for adults. Processes within the soil can release aluminium out of clay minerals. Transported into the recipient **aluminium** is a strong fish poison. If the alkalinity is close to zero within drainage waters aluminium can raise up to more than 10 mg/l (MILJØSTYRELSEN 1986). **Higher turbidity** leads to reduced weight of some fish species, because prey can hardly be detected. This is problematic especially for fish that feed on swimming organisms. **Clogged interstitial** gap systems are lost as spawning grounds. Fish avoid spawning within clogged substrates. Eggs suffocate because of insufficient current and oxygen supply in the interstitial. Ochre reduces reproduction rates of species that spawn during winter season, at plants and substrates. This is altering the composition and age structure of fish fauna.

One example of international importance is the **houting** (*Coregonus oxyrinchus*), a Flora-Fauna-Habitat species. It lives in the Wadden Sea and ascends mainly into the Vide System in autumn for spawning. It needs streams with a good current, fine spawning grounds and wintergreen plants. Its spawning period has a short duration of two to three weeks during November and December. Adhesive eggs get stuck at plants and substrate. Larvae come out in February to March and feed on drifting small animals (JENSEN et al. 2003). The houting has been mainly disappeared because free passage between Wadden Sea and river systems was reduced and marsh lands were embanked. Remove obstacles as well as reduction of ochre pollution is part of "Management Plan for the Houting" and "National forvaltningsplan for laks". Figure 2 shows as an example in how far abundance and age structure of the houting are negatively influenced by ochre pollution, which impacts life cycle of the houting as well as other biological quality parameters of the WFD negatively in several ways.

---

[3] URL: http://www.mst.dk/udgiv/publications/2000/87-7944-233-1/html/kap04_eng.htm [29.08.05]

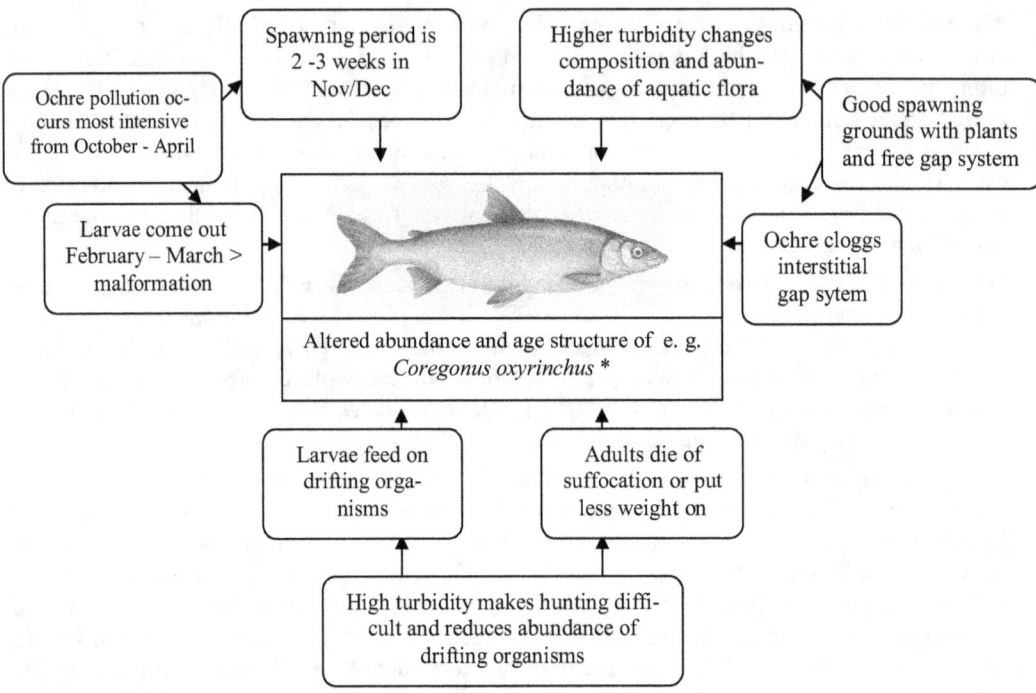

Figure 2: Influence of ochre pollution on different stages of the houting life cycle. (*Picture source: RIBE AMT & SJA 1997)

> - Verockerung beeinflusst sowohl Makrophyten als auch das Phytobenthos negativ (in Abundanz und Diversität).
> - Eine Konzentration von > 2,5 mg/l zweiwertigen Eisens hat eine biologische Gewässergütebestimmung unmöglich gemacht.
> - Gelöstes Eisen erstickt adulte Fische sowie Fischeier vor allem in Kombination mit niedrigen pH-Werten.
> - Verstärkte Trübung vermindert die Nahrungsaufnahme bei Fischen.
> - Verstopfte Kieslückensysteme verringern das Vorkommen von benthischen Organismen und Laichsubstraten.
> - Maßnahmen zur Bekämpfung der Verockerung sind Bestandteil z.B. des „Schnäpel Management Plans".

1.5 Natural abundance of iron within terrestrial and aquatic environments

Ochre pollution has its origin in the iron compounds within geology, groundwater, aquifers and soil. Anthropogenic alterations of hydrological systems cause ochre pollution within surface waters where it would not appear usually. Ochre is mostly caused by the oxidation of iron. Following aluminium, iron is one of the most important elements of the continental earth crust (4.2 %). Iron is a trace element and occasionally classified as a heavy metal. It occurs mostly within compounds with silicates, oxides, sulphides and organic material or within chelates. High contents of iron oxides are bound within granite and basaltic-gabbroid rocks (8.6 %) but also within silicate rocks and clay rocks (4.8 % or 39 g Fe/kg). Rock material that was developed by sediments and shifting sands contain less iron of approximately 19 g/kg (GRIEBLER & MÖSSLACHER 2003, SCHEFFER et al. 1998).

Iron is a main compound of minerals like goethite, hematite, siderite and others. The most important **mineral** according to ochre pollution is **pyrite (FeS₂)**. It is ubiquitous appearing together with brown coal (lignite) or as fossils and within the Gr-horizon of gley-soils (KÖLLE et al. 1983). The formation of pyrite needs **iron, sulphur, organic matter and anaerobic conditions**. During the tertiary period (65 – 2 million years ago) the climate was humid and wetlands and woods covered the landscape. These were covered by sediments when the sea level rose again (RÖCKMANN 2001). Pyrite compounds become instable under aerobic conditions and are washed out into surface waters. Although high concentrations of iron can appear at springs and wells (1-10 mg/l), within surface waters iron occurs naturally in very small amounts (µg/l). Higher concentrations are caused allochthonous. This means it has been transported from the catchments into the streams and lakes. Ochre pollution starts in the catchments area and the identification of the reasons demands knowledge of interacting pedological, hydrological, chemical and microbial processes. Oxygen, alkalinity and pH are key factors.

- Eisen kommt mit relativ hohen Gehalten in basaltisch-gabroiden Gesteinen, aber auch anderen Gesteinen sowie in verschiedenen Bodentypen und Mineralien vor.
- Pyrit ist im Zusammenhang mit der Verockerung das wichtigste Mineral, welches ubiquitär vorkommt und unter aeroben Bedingungen instabil wird.
- In Oberflächengewässern kommt Eisen natürlicherweise nur in sehr geringen Konzentrationen vor.
- Verockerung entsteht im Einzugsgebiet und wird durch ein komplexes Zusammenspiel von bodenkundlichen, hydrologischen, chemischen und mikrobiologischen Prozessen zum gewässerökologischen Problem. Sauerstoff, Alkalinität and pH-Wert sind hierbei entscheidende Schlüsselfaktoren.

1.6 Chemical iron oxidation

1.6.1 A redox-reaction determines the relation between ferrous and ferric iron

Iron can occur as dissolved ferrous iron (Fe^{2+}) and ferric iron (Fe^{3+}). Ferric iron itself is immobile. It can appear dissolved as iron hydroxides, -oxides particles or in compounds with organic matter. Both ferrous and ferric iron is a sensitive reacting pair of ions within a so called redox-reaction (BREHM & MEIJERING 1996). Equation (1) and (2) are reduction reactions if read from left to right and oxidation reactions if read from right to left. Equation (3) shows the complete redox-reaction.

The redox-potential describes the relation between the concentration of reduced and oxidized ions. It is the affinity of a substance for electrons, also called its electronegativity. The half reaction with the higher redox-potential is the oxidizing compartment (SCHEFFER et al. 1998). This reaction determines the solubility and availability of iron ions in the nature. The reaction is influenced by different factors. It is important to know that **both iron types can occur** at the same time **but in different relations to each other**. The redox-reaction is an equilibrium reaction that responses on alterations of chemical parameters and determines therefore the phenotype of ochre pollution and if organisms are stressed by toxic ferrous iron or by ochre.

Equation (1) $\qquad Fe^{2+} + 3\ H_2O = Fe(OH)_3 + 3\ H^+ + e^-$
Equation (2) $\qquad 2\ H_2O = O_2 + 4\ H^+ + 4e^-$
Equation (3) $\qquad 4\ Fe^{2+} + O_2 + 10\ H_2O \leftrightarrow 4\ Fe(OH)_3\downarrow + 8\ H^+$

1.6.2 Chemical iron oxidation and its interaction with chemical parameters

On the one hand **chemical parameters** determine the reaction velocity and consequently the distance that ferrous iron is transported downstream. On the other hand iron oxidation alters the water chemistry in a way that is disadvantageous or even toxic to organisms. In contact with water ferrous iron is hydrolyzed first (equation 4). Protons are released. The following reaction consumes oxygen and iron hydroxide is precipitated (equation 5). Both processes are stressing the ecosystem. The oxidation of ferrous iron is **lowering the oxygen concentration as well as the pH** of surface waters. Iron mostly occurs in compounds with e.g. silicates, sulphides or organic matter. Within water these iron salts are also hydrolysed. Solutions of ferrous iron salts are acid (equation 6 and 7). Iron hydroxides loose water in successive steps (equation 8) and can become very stabile and even resistant to strong acids (BREHM & MEIJERING 1996). KUNTZE (1978) mentioned also ferrihydrite (5 Fe_2O_3 * 9 H_2O) which is amorph and appears, when iron rich water is suddenly exposed to aerobic conditions.

Equation (4)	$2 Fe^{2+} + 2 OH_2 \rightarrow 2 FeOH^+ + 2 H^+$	acidification
Equation (5)	$2 FeOH^+ + ½ O_2 + OH^- + OH_2 \rightarrow 2 Fe(OH)_3$	loss of oxygen
Equation (6)	$FeA_2 + OH_2 \leftrightarrow FeOH^+ + H^+ + 2A^-$	slightly acidification
Equation (7)	$FeA_3 + 3 OH_2 \leftrightarrow Fe(OH)_3 + 3 H^+ + 3 A^-$	strong acidification
Equation (8)	$2 Fe^{3+} + 6 OH^- - (n-3) H_2O \rightarrow [2 Fe(OH)_3] \rightarrow Fe_2O_3 * n H_2O$	ageing

Under sterile conditions, at 4.5 < pH < 8 and an alkalinity of almost zero, **oxygen** and the **pH** are determining the reaction velocity. High oxygen saturation is accelerating the chemical oxidation of ferrous iron. Highest oxygen concentrations appear during autumn (November) because of lower temperatures and less biological activity. The higher the pH the higher is the reaction velocity of the chemical oxidation and the shorter is the half life of ferrous iron. The oxidation is accelerated by a **factor of 100** if the **pH is raised one unit**. At pH < 4.5 chemical iron oxidation does not take place. At pH < 7 it is slow. Drainage waters are seldom having pH < 4.5. In 1985 samples of drainage waters were taken all over Jutland. Only 5 % of the samples had a pH < 4.5. Such extreme values occur within streams close to brown coal (lignite) mines or if the alkalinity is close to zero (MILJØSTYRELSEN 1984).

According to the consequences of ochre pollution to the ecosystem, **alkalinity** is the **most important** factor. Ferrous iron is oxidized and immobilized in the soil and prevented to be washed into the streams. Half life of ferrous iron was much lower within streams of high alkalinty (REKER 1984). Within surface waters the alkalinity is represented mostly by carbonate and hydrogencarbonate. This natural buffer system is dependent on the geology. Surface waters with underlying silicate rocks have usually a low alkalinity of < 0.5 mmol/l (UHLMANN & HORN 2001). The alkalinity within South-East Jutland is naturally high (2.5 to 3.0 mmol/l) because of glacial deposits. In comparison to that alkalinity is low (0.5 to 0.7 mmol/l) in the North-West (e.g. Department Ringkøbing). Ferrous iron is kept in solution under reductive conditions caused by organic matter. Ferric iron acts as an electron acceptor here. A pH < 7, low oxygen saturations and photoreduction favour higher concentrations of ferrous iron. The average concentration of ferrous iron is 0-2 mg/l in South Jutland. Interactions between chemical parameters and iron red-ox-reaction are shown in figure 3.

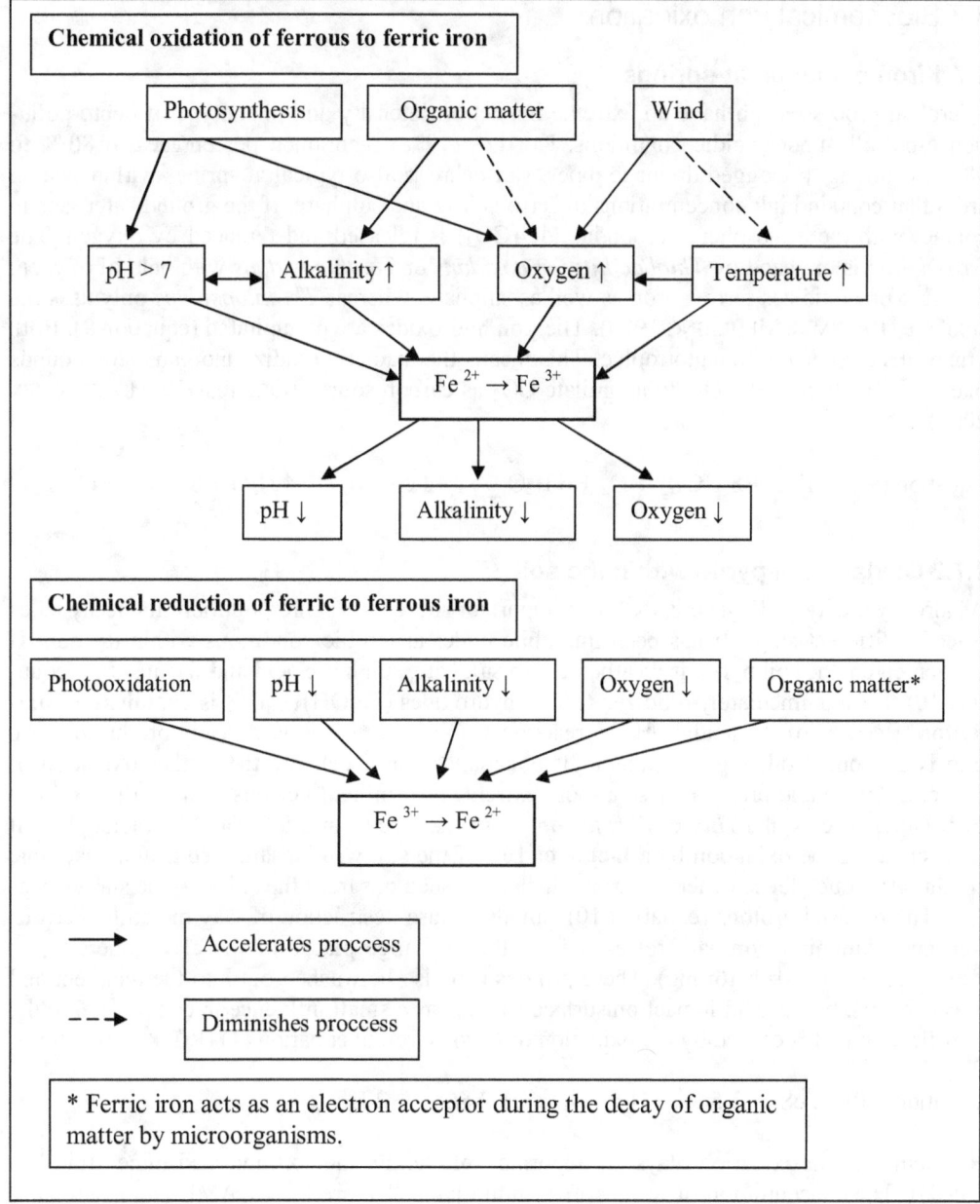

Figure 3: Interaction of physico-chemical parameters and its influence on the redox-reaction of iron.

- Die chemische Oxidation ist eine Redoxreaktion. Durch die vorherrschenden physikalisch-chemischen Bedingungen wird die Reaktionsgeschwindigkeit bestimmt und damit auch, wie weit toxisches gelöstes Eisen stromabwärts transportiert wird.
- Eisenoxidation belastet den Sauerstoffhaushalt der Gewässer und senkt den pH-Wert.
- Je höher Sauerstoffkonzentration und pH-Wert, umso kürzer ist die Halbwertszeit des gelösten zweiwertigen Eisens.
- Eine hohe Alkalinität fördert die Oxidation und damit die Immobilisation zweiwertigen Eisens im Boden bzw. dessen Sedimentation im Gewässer.

1.7 Biochemical iron oxidation

1.7.1 Iron oxidation at springs

Microbial processes can have an extreme impact on intensity and phenotype of ochre pollution especially under acidic conditions. KUNTZE (1978) mentioned percentages of 80 % to 98 % according to clogged drainage pipes. But ochre is also typical at springs within swamp areas that contain high concentrations of ferrous iron and sulphate. If the groundwater gets in contact with the atmosphere carbondioxide (CO_2) is released and replaced by oxygen. The ferrous iron is oxidized by *Thiobacillus ferrooxidans* or *Th. thiooxidans* with help of oxygen. The first one oxidizes ferrous iron as well as sulphate, whereas *Th. thiooxidans* only uses the sulphate (BREHM & MEIJERING 1996). The iron hydroxides are precipitated (equation 9). Both organisms are chemolithoautotrophic. This means that bacteria oxidize inorganic compounds like Fe^{2+}, H_2, H_2S, NH^{4+} etc. to assimilate CO_2 as carbon source (GRIEBLER & MÖSSLACHER 2003).

Equation (9) $\quad 4\ Fe^{2+} + 8\ HCO_3^- + O_2 + 6\ H_2O \rightarrow 4\ Fe(OH)_3\downarrow + 4\ H_2O + 8\ CO_2\uparrow + e^-$

1.7.2 Oxidation of pyrite within the soil

As already mentioned pyrite is the most important source for ochre pollution. Its decay takes place in different steps. It has been immobile under anaerobic conditions within the soil. If oxygen enters the soil pyrite is weathered into sulphate, sulphuric acid and ferrous iron (equation 10) that is immediately oxidized to iron hydroxides ($Fe(OH)_3$). This is **chemical autoxidation**. Because of the acidic pH the reaction velocity is very low. Already produced ferric iron is a strong oxidizing compound. Its abundance can accelerate the further oxidation of pyrite. Sulphate and protons cause a more extreme acid milieu. Ferrous iron is now oxidized by bacteria species like *Thiobacillus ferrooxidans* that occurs at 2.5 < pH < 4.5 (acidophil). It is accelerating the oxidation by a **factor of 105**. If the soil will be saturated again, anaerobic conditions reduce ferric to ferrous iron which is washed out from the soil into the surface waters. The released protons (equation 10) can also cause destruction of clay minerals. Ferrous iron and aluminium ions are released from the exchange places of the clay minerals (sekundäre Tonmineralzerstörung). These protons can also be washed out into the recipient and lower the pH, but its acid impact on surface waters has a small influence in comparison to the acidification that is caused by the oxidation of ferrous iron in equation (4) (RÖCKMANN 2001).

Equation (10) $\quad FeS_2 + 3.5\ O_2 + H_2O \rightarrow Fe^{2+} + 2\ SO_4^{2-} + 2\ H^+$

Biochemical iron oxidation plays an important role within the soil and acid mine drainages (AMD). The concentration of ferrous iron within lakes that are fed by AMD can reach more than 100 mg/l (SYMADER 2004). Figure 4 shows a tributary of the river Timm where ferrous iron concentrations are approximately 240 mg/l (RINGKØBING COUNTY 1995). Biochemical and chemical processes are hard to distinguish at neutral pH. The influence of other bacteria has been supposed to be less intensive in comparison to the chemical oxidation at high pH (REKER 1984). Nevertheless *Gallionella ferruginea* are the best known bacteria according to ochre clogged drainage pipes and wells. Iron oxidizing organisms and bacteria that occur at neutral pH will be summarized in table 1. Most of them are **gradient organisms** that live at the border between aerobic and anaerobic conditions.

Figure 4: Acid Mine Drainage within Ringkøbing County, Denmark. (Picture: H. Prange, 05/2005)

Table 1: Summary of iron oxidizing bacteria and organisms

Genus or Spezies	Metabolic activity	Abundance at ...	Literatur
Thiobacillus ferrooxidans	Chemolithoautotrophic, oxidizes ferrous iron and sulphate	2.5 < pH < 4.5; within pyrite-rich soils, acid mine drainage	KUNTZE 1978 BREHM 1996
Thiobacillus thiooxidans	Chemolithoautotrophic, oxidizes iron only	Acidic pH	BREHM & MEIJERING 1996
Gallionella e.g. G. ferruginea	Prefers pyritemonosulphate (FeS) within soils > delivers iron more constantly in comparison to pyrite, causes denitrification	Neutral pH, 1-5 °C, waters with ferrous iron concentrations of < 12 mg/l	KUNTZE 1978 EGGELSMANN 1973 KOROM 1992 POTT & REMY 2000
Leptothrix ochracea (free swimming filaments)	Facultative chemolithoautotrophic, uses organic compounds, oxidizes iron and manganese	Neutral pH	RHEINHEIMER 1991
Crenothrix polyspora (sessil filaments)	Facultative chemolithoautotrophic, use organic compounds, oxidizes iron and manganese	Neutral pH	RHEINHEIMER 1991
Thiothrix spec.	Oxidizes H_2S, iron oxides are produced facultative	Neutral pH	KUNTZE 1978
Heterotrophic bacteria and organisms (e.g. *Pseudomonas, Vibrio, Flavobacterium, Bacillus, Arthobacter, Nocardia*)	Organic carbon is used for assimilation, use N-, C- or organic compounds as energy source, iron is not used chemolithoautrophically	Neutral pH	RHEINHEIMER 1991 GRIEBLER & MÖSSLACHER 2003
Thiobacillus denitrificans	Uses the oxygen of nitrate under anaerobic conditions	Neutral/high pH	KÖLLE et al. 1983 POTT & REMY 2000

- Verockerung tritt natürlicherweise an Quellen auf.
- Im Boden und bei pH 2,5 bis 4,5 hat die biochemische Eisenoxidation durch Bakterien der Gattung *Thiobacillus* eine große Bedeutung.
- Im neutralen Bereich können Bakterien der Gattung *Gallionella* spec. und andere Organismen eine wichtige Rolle spielen.

1.7.3 Sediments and resuspension within standing waters

Referring to UHLMANN & HORN (2001) the maximum depth, surface area, maximum effective fetch (length), volume and retention time determine budget of materials (Stoffhaushalt) within standing waters. What benefits the oxidation of ferrous iron is not automatically good for the sedimentation of ochre. The **surface-volume-ratio** is altered depending on the season and the shape of lakes. A big surface-volume-ratio (shallow water body) would benefit oxidation of dissolved iron as well as the installation of vegetation. But shallow ochre lakes are also liable to resuspension due to wind disturbance. If the whole water body is mixed, lakes are so called "holomictic". Consequently resuspension plays an important role within flat lakes. A permanent sedimentation would prevent ochre to be transported out of the lake. Resuspension occurs due to wind disturbance but also as a complicated interaction between sediment and water body.

During summer season the **ratio of sediment surface and water body** is higher. At the same time discharges and consequently water exchange are lower. An increased biological activity influences the oxygen balance and release of nutrients is more extremely (UHLMANN & HORN 2001). Oxidized iron within the sediments would be released into the water because of an anaerobic layer above the sediment. **Layering** can also occur because of the entry of groundwater that is colder or rich in salts (thermo- or chemocline layering). A layer of water with a high density can possibly prevent resuspension due to wind disturbance. Table 2 summarizes the morphometric influences on resuspending processes.

Table 2: Summary of interactions between morphometry, layering (thermo- and chemocline) and iron oxidation or reduction respectively

Parameter	Fe^{2+} Oxidation	Fe^{3+} Sedimentation
Large surface (high surface-volume-ratio)	Good for oxidation due to wind disturbance, but can accelerate photo reduction	Resuspension due to wind disturbance
Small surface (low surface-volume-ratio)	Less oxidation but also less photo oxidation	Less resuspension
Shallow water body benefits growth of vegetation	Good for oxidation (chemical and biochemical)	Offers wind shelter but is also liable to clogging and development of channels
Eventual layering	Can cause resuspension due to anaerobic conditions	Can maybe prevent resuspension

- Sowohl das Oberflächen-Volumen-Verhältnis, als auch das Sediment-Volumen-Verhältnis kann die Umsetzungs- und Freisetzungsprozesse der Eisenverbindungen maßgeblich beeinflussen.
- In flachen holomiktischen Wasserkörpern kann es unter Umständen zur Ausbildung von thermo- oder chemoklinen Schichten kommen, die sowohl die Resuspension von zweiwertigen Eisen verursachen, als auch die Resuspension von bereits sedimentierten Eisenverbindungen unterbinden könnten.

1.8 Different reasons of ochre pollution

1.8.1 Anthropogenic reasons for ochre pollution

Two main reasons of ochre pollution should be emphasized. **Drainage** means to drain adhesive, impounded or groundwater. Drainage of impounded or adhesive water is improved by a better soil structure due to the incorporation of lime and pipe less drainage. Groundwater needs to be drained by pipe systems and ditches. The depths vary between 0.4 to 1.2 m depending on soil type and land use. The other reason is the regulation and maintenance of streams by mow baskets (**intensive maintenance**). Complete removal of vegetation, excavation of sediments and stones as well as straightened channels lead to higher velocities (Figure 5). Consequently channels are deepened and groundwater tables within the adjacent aquifers are lowered. This is especially a problem at sandy soils. Both measures lead to a change from water saturated to aerobic conditions in the soil. Former stabile iron forms are transported from soil into the recipients now.

The ochre pollution caused by **acid mine drainages** will be treated as special case here. Also input of nitrate fertilizers can enhance concentrations of ferrous iron within groundwater. During the bacterial catalysed reaction by the facultative aerobic *Thiobacillus denitrifibcans* pyrite is oxidized. While denitrification is a positive effect concentrations of ferrous iron and sulphate are increasing (KÖLLE et al. 1983). **Anthropogenic caused acidification** can have also an impact, especially on the destruction of clay minerals within soils of low buffer capacity and pH < 4.5.

Figure 5: Maintained ditch close to the village Wietzen (Lower Saxony). (Picture: H. Prange, 03/2005)

1.8.2 Allochthone and autochthone ochre pollution

Different investigations showed that the iron output of freshly drained soils declines **after two to three years** (occasionally after five years). The pyrite is completely oxidized and ferrous iron is completely released and washed out. This is **autochthone** ochre pollution. In comparison to that ferrous iron can be transported by groundwater over far distances. Depending on the size of groundwater body and aquifers it is not possible to estimate intensity and duration of the ochre pollution (KUNTZE 1978).

It is hard to distinguish where ochre appears naturally and where it is caused anthropogenic. Straightened ditches or artificial recipients can cut through a groundwater body. This enhances entry of iron rich groundwater into surface waters. Most areas had been drained during the 1980's in Germany. In Denmark draining agricultural land is not granted anymore since 1990. With an **average duration of two to three years** recent ochre pollution can not be autochthone anymore. But if the groundwater level or the oxidizing horizon is constantly lowered, non-oxidized pyrite will be exposed to aerobic conditions and causes autochthon ochre pollution.

> - Verockerung wurde und wird hauptsächlich anthropogen verursacht durch die Drainage von Grundwasser sowie die intensive Gewässerunterhaltung.
> - Da die autochthone Verockerung zwei bis drei Jahre dauert, die meisten Flächen aber bis Anfang der 1980er Jahre dräniert worden sind, wird es sich bei aktuell auftretender Verockerung um durch Grundwasserzutritt verursachte (allochthone) Verockerung handeln.
> - Durch das Absinken des Grundwasserspiegels, z.B. auf Grund intensiver Gewässerunterhaltung, werden jedoch immer wieder neue Erdschichten und das darin enthaltene Pyrit der Oxidation ausgesetzt.

1.9 Aim of investigations

So far it has been reviewed that ochre pollution has its origin in natural occurring iron compounds like pyrite that were developed during thousands of years. A complicated process based on interactions between chemistry and microbiology, which has been intensified or is even appearing because of anthropogenic alterations of the landscape and hydrological processes. A mostly seasonal visible problem that has been identified in different literature sources as threat to biological quality parameters of the WFD e.g. fish, benthic invertebrates and macrophytes.

Ochre pollution has been tackled as an obvious problem in the aquatic environment in Jutland for more than ten years now. This dissertation work aimed to describe methods and summarize already abundant experience of ochre abatement measures in South Jutland, mainly the SJA. The investigation of three ochre lakes will also show what parameters cause different cleaning efficiencies and prove that ochre pollution is a problem with a negative long-term impact on the aquatic environment. Finally other measures (e.g. restoration of wetlands and altered maintenance) to combat ochre pollution will be described to provide an overview of different possibilities of which the instalment of ochre lakes is only one.

> - Durch den anthropogenen Einfluss werden natürlich vorkommende, immobile Eisenverbindungen freigesetzt.
> - Verockerung ist ein zumeist periodisch auftretendes Problem, dessen negative Einwirkungen auf die biologischen Qualitätskomponenten der Wasserrahmenrichtlinie bereits in verschiedenen Literaturstellen beschrieben wurde und das seit mehr als zehn Jahren in Jütland bekämpft wird.
> - Ziel dieser Arbeit ist es, die in Jütland angewandten Lösungsansätze aufzuzeigen und bereits bestehendes Wissen zusammenzufassen.
> - Neben den möglichen Gründen für unterschiedliche Reinigungseffizienzen von drei Ockerseen soll des Weiteren klar werden, dass die Verockerung ein Problem mit negativer Langzeitauswirkung auf das aquatische Ökosystem ist.
> - Anschließend werden weitere Bekämpfungsmaßnahmen wie z.B. die Wiedervernässung beschrieben, um einen Überblick über mögliche Gegenmaßnahmen zu geben, von welchen die Anlage der Ockerseen nur eine ist.

2. Materials and methods

2.1 General approach

A first excursion to the SJA took place on 15th March 2005. Different lakes were visited and first information provided by the River Section of the SJA. In advance of the ten weeks stay in South Jutland (Denmark) literature research was done. From mid of April to the end of June it was one main goal to **review and deepen the results** of the project report that was written during the 7th semester of the International Degree Course Environmental Biology, University of Applied Sciences Bremen. Literature and file research was done mostly in Danish and English. Information was also gained during interviews with staff from different sections of the SJA (Groundwater Section, Wastewater Section and River Section). Additionally the Ringkøbing County, the Holstebro Municipality and the Herning Municipality were visited on 10th May, 1st and 2nd of June. Question leaflets were sent in advance. The **site visits** enabled the enhancement of knowledge about different origins and abatement measures of ochre pollution. Reports, sketches and investigation results as well as notes and pictures of the site visits are described within chapter 3.2 and 3.3.

2.2 Detailed investigations of three ochre lakes in the SJA

2.2.1 Structure of the investigation of ochre lakes

The investigation started with the extraction of data of 26 ochre lakes and a maximum period of twelve years from the SJA data bases. Water samples (taken all two months) are investigated for iron concentrations by SJA staff and based on the methods as described under "Dansk Standardisieringsgrad 219 – Bestemmelser af jern" (DS 219). Discharge, oxygen concentration, pH and temperature have been measured all two months. Additional data acquisition will be explained within the following chapters.

The first step was to get an overview about available data. Information about the installation and construction of the lakes were gained from different files, e.g. project proposals, reports, ArcView© shapes and collected in data sheets. Different lakes were visited. The investigation was finally emphasized on three lakes. The decision was based on:

- type of lake

- special characteristics e.g. attempts of improvements during the next autumn

- actuality of data and sufficient amount of data

The investigation of the efficiency was subdivided into three main topics as shown in table 3. The **first** part aims to identify and explain the origin of the ochre pollution. The **second** part is concentrated on the lake itself. It should identify or at least give a hint on possible negative influences and ideas for improvements. **Finally** the influence of the lakes instalment on the ecosystem should be considered.

Table 3: Structure of investigation of ochre lakes

1) What happens upstream?	Catchments size
	Discharges
Is the lake reached by low/high amounts of dissolved/total iron	Soil types
	Kind of ochre source
	Development of ochre pollution, phenotype
	Correlation with chemical parameters
2) What happens in the lake itself?	Concentration of dissolved iron at the inlet
	Concentration of dissolved iron at the outlet
Acceleration or decrease of iron oxidation/ochre sedimentation within the lake	Load of iron total at inlet (based on median, winter season)
	Load of iron total at outlet (based on median, winter season)
	Cleaning efficiency (loads of iron total)
	Vegetation
	Deposition
	Retention time
	Surface-volume-ratio
	Appearance of channels/islands
	Alteration of chemical parameters
3) Influence of the lake on the watercourse/ecosystem	Stream target (Gewässerzielsetzung)
	Free passage for fish and benthic invertebrates
	Chemical water quality

- Die Untersuchung der Ockerseen wurde unterteilt in drei Bereiche.
- Es wurde untersucht, welche Ursachen die Verockerung im Einzugsgebiet hat und welche Eisenfrachten den See erreichen.
- Im zweiten Schritt wurde untersucht, was die Effizienz des Sees beeinflussen könnte (z.B. Vegetation, reduziertes Volumen und Retentionszeiten, Stromrinnenbildung).
- Zusätzlich wurde die Veränderung physikalisch-chemischer Parameter durch den See untersucht und betrachtet, in wie weit die Gewässerzielsetzung bisher erfüllt wurde.

2.2.2 Collection of additional background information

First of all different information were gained from SJA files, e.g. project proposals, modelling reports, regulatives, Arc View shapes etc. Information were collected within data sheets and finally summarized.

- Hintergrundinformationen wurden aus SJA Unterlagen (inkl. Arc View shapes) zusammen getragen.

2.2.3 Investigation of iron concentrations and influence of discharge

Data were got out of the SJA databases (STOQ) and transferred into MS EXCEL$^{©}$. In general winter season was defined from October to April and summer season from May to September. Averages and standard deviation were calculated for all data of concentration and following for the medians of winter and summer seasons. Here it was investigated how high the concentrations have been and if threshold limits are still exceeded or not. Finally it is shown in how far the discharge is correlated with iron concentrations at the inlet. Also correlation between ferrous iron and precipitation per month and per day was investigated for each lake.

> - Für die Konzentrationen am Ein- und Auslauf des Sees wurde der Durchschnitt aller Werte und Mediane für die Winter- und Sommersaison berechnet. Dies ist notwendig um zu ermitteln, in wie weit Grenzwerte überschritten werden.
> - Korrelationen zwischen der Konzentration des gelösten Eisens am Einlauf mit dem Abfluss und dem Niederschlag wurden ebenfalls untersucht.

2.2.4 Investigation of iron loads

Loads of dissolved and total iron have also been calculated in the same way as the concentrations for each ochre lake. This gives an imagination of kg/day or t/year that were prevented to pollute the watercourse downstream the lake. The calculation of the loads was based on the medians in kg/day. These have been multiplied with 213 days per winter season and 152 days per summer season. Data are enclosed in the appendix. Tables that were necessary for the calculation but were not mentioned in the results can be found on the attached CD-ROM.

> - Die Kalkulation der Frachten in kg/Tag oder t/Jahr gibt eine Vorstellung von der Belastung, der die Gewässer ausgesetzt sind, und welche Eisenmengen von den Seen abgefangen werden.

2.2.5 Investigation of abundance and distribution of vegetation and turbidity

Hvirlå and Landeby Bæk were subdivided into three or four compartments in which the cover of abundant water plant species was estimated. The vegetation at Løgumkloster Bæk Ochre Lake was not investigated in detail. The turbidity was determined with a Secchi-disk at the deepest part of Hvirlå and Landeby Bæk Ochre Lakes.

> - Es erfolgte eine prozentuale Schätzung der Pflanzendecke im Hvirlå and Landeby Bæk Ockersee. Außerdem wurde die Sichttiefe gemessen.

2.2.6 Investigation of deposition, retention time and surface-volume-ratio
2.2.6.1 Hvirlå Ochre Lake

Discharges have been measured over the last twelve years. The volume of this type of lake is relatively constant. Therefore it was possible to calculate the theoretical retention times for every date by division of the theoretical volume through the discharges and correlate it with the cleaning efficiencies before and after the enlargement. The extension of the Hvirlå Ochre Lake before and after the enhancement can be seen on the two sketches included in the appendix. Investigations took place on 15th June. A transect was laid across each compartment and end points were digitalized. The distance between water surface and bottom was measured every one to four meter. Following the distance was substracted from the actual height of water table to get the height of ochre deposition in m DNN (Dansk Normal Null). The average of level values at in- and outlet before and after the measurements were calculated. Cross-sections of each compartment were produced in MS EXCEL.

> - Auf Grund der guten Datenlage konnten für den Hvirlå Ocker See die theoretische Aufenthaltszeit berechnet und mit den Reinigungseffizienzen, vor und nach der Erweiterung des Sees, korreliert werden.
> - Die Sedimentdicke wurde in drei Transekten ermittelt.

2.2.6.2 Løgumkloster Bæk Ochre Lake

The first step was to calculate the exact theoretical volume without any deposits. Therefore the sketch of the ochre lake was subdivided into eight compartments (sketch enclosed in the appendix). Typical cross-sections were put into every compartment and data were transferred into MS EXCEL. The volume was calculated for each compartment up to 10.45 m DNN, 10.75 m DNN and 11.00 m DNN. This was necessary because the volume of Løgumkloster Bæk Ochre Lake is increasing with higher discharges.

In a second step the height of deposition had to be found out to calculate the actual volume. During a site visit on 24th of June the height of deposits was measured. The volume of the deposits was calculated in the same way as explained above and subtracted from the theoretical volume. Usually retention time can be calculated by dividing the volume through the discharges. The level values were not available. Discharge has been measured a few meters downstream at Løgumkloster Marketsplads (Station-No.: 4032090). A **Q/Q-correlation** had to be done. This means Løgumkloster Bæk was compared with another station which catchments size and precipitation is nearly the same. At Landeby Bæk (Station-No.: 40.09) discharges are regularly measured. Five graphics, that show the comparison of measured and calculated data of both measuring stations as well as the trend line with a good quality coefficient of $r^2 = 0,98$, are included in the appendix. The data of Landeby Bæk reference station were transfered to MS EXCEL, sorted and the 50 % and the 95 % percentiles could be calculated. Because the catchments area of the lakes outlet is 4 km² (14 %) smaller than that of Løgumkloster Marketsplads this had to be included into the calculations, too. (Data on CD-ROM)

Finally the 50 % and the 95 % percentile could be used to calculate the retention times at the three different heights. In comparison to the Hvirlå Ochre Lake it was not possible to investigate the development of the retention time more detailed. The main interest was to calculate in how far an enlargement of the volume would increase the retention time. Finally data of cross sections that were already produced for the calculation of the volume were also used to calculate the surface-volume-ratio.

- Auf Grund der Planungsskizze vom Løgumkloster Bæk Ocker See konnte das theoretische Volumen des Sees bei unterschiedlichen Wasserständen berechnet werden. Dies war notwendig, da der See mit steigendem Zufluss an Volumen gewinnt.
- Da die Pegelstände der vergangenen Jahre nicht ermittelt worden waren, wurden 50 % und 95 %-Percentil mit Hilfe einer Q/Q-Korrelation berechnet, um hiermit die evtl. Verlängerung der Retentionszeit durch die Dammerhöhung berechnen zu können.
- Anschließend wurden die zuvor erstellten Querschnitte in den Seeabschnitten zur Beurteilung der Veränderung des Oberflächen-Volumen-Verhältnisses herangezogen.

2.2.6.3 Landeby Bæk Ochre Lake

Four transects were positioned across the lake (sketch is enclosed in the appendix). The distance between water table and bottom was measured every one to four metres. Because there were no levels, the height in m DNN had to be calculated by using the height that is mentioned in the regulative files. The height of the water table was measured at the inlet and added to this value. Data were fed into MS EXCEL to produce cross sections and the actual volume could be calculated. Additionally some sediment cores were taken. Oxygen concentration, pH and temperature were measured within different depths at the deepest part of the lake.

> - Auch im Landeby Bæk Ockersee wurde die Sedimentdicke ermittelt, um reduziertes Volumen und Retentionszeit berechnen zu können.
> - Mit einem Sedimentprobennehmer konnten Einblicke in das Sedimentprofil gewonnen werden.
> - An der tiefsten Stelle des Ockersees wurde ein Tiefenprofil von Sauerstoff, pH und Temperatur erstellt.

2.2.7 Determine discharge within the vegetation channel at Hvirlå Ochre Lake

This investigation took only place at the Hvirlå Ochre Lake. On 24th June the discharge [l/s] was measured at in- and outlet and within the vegetation channel that had been detected on 15th June to investigate in how far it is decreasing the cleaning efficiency. A rope had to be put again across the second compartment to ensure a relative stabile position of the boat. Finally data were fed into HYDROS (PC software) to produce the discharge cross sections.

> - Da am 15. Juni eine große, einzelne Stromrinne im Hvirlå Ockersee festgestellt wurde, sollte anschließend der Abfluss in dieser gemessen und ihr Einfluss auf die Reinigungsleistung abgeschätzt werden.

2.2.8 Data investigation of oxygen, pH, alkalinity and temperature

Average, median and standard deviation were calculated for all chemical parameters. It was investigated in how far the lake has altered the chemical parameters in a way that could benefit the iron oxidation and therefore cleaning efficiency (e.g. higher pH or oxygen concentration). It was also examined if the lake alters the chemical parameters in a disadvantageous way for the ecosystem (e.g. temperatures or oxygen household). Ferrous iron concentration and chemical parameters at the inlet were correlated to investigate in how far the chemical conditions upstream influence the iron load that reaches the lakes. Figures that show a correlation were included in the results. Alkalinity was only measured during the 1990´s. Because Landeby Bæk is one of the newest lakes data of alkalinity concentrations are not available.

> - Es wurde untersucht, in wie fern die durch den Ockersee veränderten physikalisch-chemischen Parameter die Eisenoxidation verstärkt haben könnten.
> - Außerdem wurde beurteilt, ob sich physikalisch-chemische Parameter stark und damit negativ für den Unterlauf verändert haben könnten.
> - Korrelationen physikalisch-chemischer Parameter mit der Konzentration des gelösten Eisens am Einlauf der Seen sollten Aufschluss über den Einfluss des Einzugsgebietes geben.

2.2.9 Equipment for the investigations within the field

Following materials have been used during the investigations at Hvirlå and Landeby Bæk Ochre Lake:

Table 4: Equipment and materials for investigations of ochre lakes

Investigations at Hvirlå Ochre Lake	Investigations at Landeby Bæk Ochre Lake
Equipment for discharge measurement: G.M.I. Geological & Marine Instrumentation ApS	Sediment samplers Water sampler Measuring instrument for oxygen and pH: WTW, Multiline P3; pH/Oxi-Meter
Prepared mapping sheets Boat GPS Ropes with 1 m marks Measuring rod Iron hooks Rake and bucket Secchi-disk Digicam	

Anmerkung: Bei Interesse können die Daten des Anhangs als CD-ROM angefordert werden.

3. Results

3.1 Abatement measures and strategies from Denmark

3.1.1 Identification of ochre potential areas and threshold values

In 1981 a notification was published (the Drainage and Irrigation Subsidies Act) which demanded the consideration and calculation of costs of ochre abatement measures while applying for drainage subsidies. Based on this it was decided to investigate the impact of ochre on stream ecology. Investigations and experiments done by the Danish Land Development Service (Hedelseskabet) took place from **1981 to 1984**. The target was to map ochre potential areas, define threshold values that protect aquatic flora and fauna and develop measures to remove iron from drainage waters (CHRISTENSEN AND OLESEN 1985). The **Ochre Act** itself was passed in **1985** (Law No. 180 passed at the 8th May 1985). It requires that drainage projects within ochre potential areas have to be allowed by the responsible County.[4]

Ochre potential areas were identified at **historical maps** from 1921 – 1929. Frames were drawn around areas of bogs and wet meadows. Additional the content of sulphate was determined to classify ochre potential areas into I (red), II (green), III (blue). Round about 300.000 ha in Jutland were identified as ochre potential, which is 10 % of the total area. Ochre pollution occurs mostly within the Countys of Ribe, Ringkøbing and South Jutland. In 1986 approximately 28 % of watercourses were estimated to be polluted because of their location within or downstream ochre potential areas.[5] In the SJA approximately 450 km² were identified as **ochre potential** of which 280 km² were defined as class I areas (SJA a 2002). During a mapping **revision in 1992/93** approximately 12.000 ha were proposed to be removed from the map and 9.000 ha were added to the map.[6] If drainage projects within the green or blue areas are planed the pyrite content has to be determined. **Threshold limits for pyrite contents** were needed on which the decision is based whether drainage can threaten the recipients or not. Therefore investigation of pyrite and iron contents within soil and drainage water took place in **1985** all over Jutland. The Environmental project No. 78 (MILJØSTYRELSEN 1986) describes methods and results for 60 locations. The aim was to identify threshold values of pyrite within mineral and organic soils. The investigation of soil samples included the pH (Rt = Grad der Versauerung), bases (alkalinity), pyrite and free pyrite. Drainage waters were investigated for pH, alkalinity, ferrous iron and total iron, sulphate (SO_4^{2-}) and aluminium. Samples were taken during **October and two to three years after the drainage** took place, because ochre pollution is most intensive than. Investigation results led to the threshold limits shown in table 5.

Table 5: Administrative threshold limits (MILJØSTYRELSEN 1986)

Dry matter	Rt > 4.5		Rt < 4.5	
	Pyrite	Free pyrite	Pyrite	Free pyrite
Organic soil (rich in organic material) > 10 %	1.5 %	0.5 %	0.3 %	0.0 %
Mineral soil (poor in organic material) < 10 %	0.5 %	0.2 %	0.1 %	0.0 %

[4,5,6] URL: http://www.mst.dk/udgiv/publications/2000/87-7944-233-1/html/kap04_eng.htm [29.0805]

Pyrite and consequently ochre pollution can be a problem within areas under influence of brackish water like **marshes and organic soils** as swamps. Within the County of Ribe for example ochre pollution appears mostly in marsh and bog areas. Sandy soils contain also pyrite but in lower concentrations. Within the Northern and Western parts of Jutland alkalinity is low and therefore the content of free pyrite in the soil is high with ~ 0.11 % (MILJØSTYRELSEN 1986). The alkalinity within soils is determining the amount of iron that is entering the recipients. If alkalinity is high, the released ferrous iron is oxidized immediately and prevented to enter the stream.

The important point is the **change of chemical conditions** between different soil types or aquifers. KUNTZE (1978) mentioned transitional zones between bog and marshes as well as bogs and valleys as extreme ochre potential areas and bog gley soil (Moorgley), bog marsh (Moormarsch) and fen with coastal marsh sediments (durchschlickte Niedermoore) as extreme ochre potential soil types according to **autochthone ochre pollution**. Because of the strong **ground water influence** within gley soils, these and related types are high ochre potential (Anmoorgley, Naßgley, Podzolgley, Auengley). Suddenly changes of chemical conditions are determining location and intensity of ochre pollution. If iron rich groundwater from geologically aquifers enters a valley bottom rich in lime, ferrous iron is oxidized and fixed in the soil (eisenschüssiger Boden). If the valley is rich in organic matter and therefore characterized by reductive conditions, ferrous iron will be transported into the recipient (KUNTZE 1978).

3.1.2 Approval of drainage projects and stream targets

Although the amount of drainage projects decreased, administration should be explained here. Drainage of adhesive or impounded water does not need an external expertise from the Danish Land Development Service. If a farmer wants to **drain groundwater** within the green or blue classified areas, the pyrite content has to be determined. If the threshold limits (table 5) are exceeded the farmer can decide if he wants to do a so called **"model calculation"**, in which would be calculated how much iron will be transported into the recipient. If the groundwater contains > 5 mg/l ferrous iron also a model calculation should be done. Because of huge fluctuations of the pyrite and iron output of mineral soils investigations should be done just in case if the recipients are A or B streams (CHRISTENSEN & MARCUS 1998). The farmer has to pay the investigation and it is possible that the drainage project will not be allowed by the responsible county (SJA WASTEWATER SECTION, CLAUSEN personal notification 2005). But only 8 % of applications were refused, while in **79 % of the cases drainage was allowed**. Ochre abatement measures are financed by the Danish Environmental Protection Agency (DEPA = Miljøstyrelsen). In case of a prohibition the farmer gets compensation, because he was not allowed to enhance the economical value of the land. The farmer is responsible for the maintenance and the local County controls the cleaning efficiency for at least two years.

The "guidelines for recipient quality planning" No.1 was published by the DEPA in 1983 (table 6, column 1). These are included in the County's regional plan and act as "guidelines for the **quality and utilisation of watercourses** in relation to an overall assessment of their development".[7] Based on the Ochre Act, the Stream Law and the Nature Protection Law threshold limits for ferrous iron concentrations were defined (table 6, column 2). Total iron concentrations should not exceed 1.5 mg/l (SJA a 2002). Also based on the Ochre Act are threshold limits of increased ferrous iron caused by drainage projects (table 6, column 3). Several streams are classified with an additional (F), e.g. B2 (F). This **double target** means that this

---
[7] URL: http://www.mst.dk/udgiv/publications/2000/87-7944-233-1/html/kap04_eng.htm [29.0805]

watercourse would usually fulfil the target within the recent planning period but is polluted by ochre. Theoretically ochre pollution could decrease (autochthone pollution) or ochre abatement measures need to be applied. **Stream targets** determine maintenance and administration and the status of the streams are not allowed to become worse according to fauna and flora (CHRISTENSEN & MARCUS 1998).

The DEPA provides also funds for ochre abatement projects outside the identified areas. The Ochre Act has caused a reduction or at least stabilization of ochre pollution within ochre potential areas. The problem is that **ochre pollution occurs also outside these areas** as a result of drainage and regulation that took place years ago, while the Ochre Act administrates recent drainage projects (which numbers have decreased). Ochre pollution within these streams is also caused by intensive or illegal maintenance.[8]

Table 6: Threshold limits for the administration of the Ochre Act (Source: CHRISTENSEN & MARCUS 1998)

Rezipient Type	Threshold for the content of ferrous iron (average during winter season) [1]	Maximum increase of ferrous iron permitting drainage projects
A Watercourse of special scientific interest	0.2 mg/l	0.0 mg/l
B1 Spawning and nursery area for salmonoids	0.2 mg/l	0.1 mg/l
B2 Salmonoids waters (Lachsfischgewässer)	0.2 mg/l	0.1 mg/l
B3 Cyprinoids waters (Karpfenfischgewässer)	0.5 mg/l	0.1 - 0.2 mg/l
C Watercourses used only to convey water	No threshold limit [2]	Individual [2]
D Watercourses that are influenced by waste waters	No threshold limit [2]	0.1 mg/l
E Watercourses influenced by water intake (Wasserentnahme)	No threshold limit [2]	0.1 mg/l
F Watercourses influenced by ochre	No threshold limit	0.0 mg/l [4]
Water courses without any targets	No threshold limit [3]	Individual [2) und 3)]

1) In average measured concentration of ferrous iron during the winter period from October to April
2) Evaluation based also on watercourses that are located up- and downstream (Bewertung auch nach stromauf- und abwärtsliegenden Gewässerabschnitten)
3) The recent existing flora and fauna is not allowed to be reduced (Das existierende Pflanzen- und Tierleben darf nicht verringert werden)
4) Short-term discharges of groundwater can be accepted (Kurzzeitige Ausleitungen von Grundwasser können akzeptiert werden)

---

[8] URL: http://www.mst.dk/udgiv/publications/2000/87-7944-233-1/html/kap04_eng.htm [29.0805]

- Verockerungsgefährdete Areale sind anhand ehemaliger Feuchtgebiete auf historischen Karten plus zusätzliche Untersuchungen der Zersetzungsprodukte des Pyrits identifiziert worden.
- Es gibt Pyrit-Grenzwerte für mineralische und organische Böden in Abhängigkeit der Bodenreaktion (Alkalinität).
- Anhand der Grenzwerte wird entschieden, ob eine Modellberechnung zur Einschätzung der Verockerungsgefahr durch Drainage notwendig ist.
- U. a. auf Grund des Ockergesetzes von 1985 sind auch Grenzwerte für den Eintrag von gelöstem Eisen in die verschiedenen Gewässertypen festgesetzt worden.
- Das Ocker-Gesetz befasst sich mit der Administration von Dränprojekten, deren Zahl jedoch abgenommen hat. Verockerung tritt aber auch im verstärkten Maße außerhalb der verockerungsgefährdeten Areale durch intensive oder illegale Unterhaltung auf.

3.1.3 Where to start – the priority list of the Sønderjyllands Amt

In 1994 the **first Ochre Action Plan** was implemented. In advance iron contents in several streams were investigated to identify main sources with the highest iron input. The calculation of priority points was based on a cost-benefit-analysis and did not consider any ecological values of the streams. Its calculation is explained in the info box below. **Point sources** and **diffuse sources** were distinguished. The first ones are small tributaries that can be clearly identified. Most point sources were identified within the Vide system. Diffuse source means that groundwater enters the stream from the embankment. Calculated amounts of iron from different kinds of sources are shown in table 7. The most ochre polluted stream was the Hvirlå.

Info Box: Calculation of priority points

- Discharge [l/s] and the concentration of iron total were measured.
- The stream stretches with the most priority points are on the top of the list.
- Priority points = $\dfrac{A}{B * C * D}$ = $\dfrac{kg}{km * kr * 0.75}$

- A = iron load per winter season [kg/day]
- B = Length of the polluted stretch downstream the ochre abatement lake
- C = estimated costs (in Danish Kronen) for a lake with a retention time of eight hours
- D = the effect for the recipient, means in how far the pollution would be reduced, e.g. by a factor of 0.75

Table 7: Extension of ochre pollution in the SJA (CLAUSEN 1994)

Type of source	Number or extension of sources	Average iron load [kg/day] during winter season
Point sources	18	2884
11 diffuse sources within ochre potential valleys	42 km 630 km² valley area	3349
Diffuse sources within other (ochre potential) valleys	166 km 2500 km² valley area	

3.1.4 Review of ochre projects

Five years later (in 1999) ochre lakes had been established downstream **12** of the 18 point sources. Eight kilometres downstream the Hvirlå Ochre Lake a higher abundance of benthic organisms was observed, where a **biological determination of the water quality** (biologische Gewässergütebestimmung) was not possible before because of the extreme ochre pollution. **Two** of the eleven diffuse sources were improved by lakes. 26 km of the 166 km within ochre potential areas were improved by raising water tables (costs: 1.929.200 €). Costs of the **14 lakes** were in total 798.000 € (OKKERGRUPPEN 1999). 75 to 50 % of the ochre projects have been financed by the DEPA and later by the Danish Forest and Nature Agency. The Ochre Action Plan was reviewed again in 2001. **Data of the last 20 years** were investigated. The most important sources are **permanent.**

- Ockerhandlungspläne gibt es seit 1994 im SJA. Diese enthalten Prioritätslisten der belasteten Gewässer, welche nach und nach abgearbeitet werden.
- Die stärksten Quellen sind permanent vorhanden.
- Die meisten und günstigsten Projekte sind Ockerseen.

3.2 Parameters that influence instalment and maintenance of ochre lakes

3.2.1 Definition of the term "ochre lake"

Ochre lakes are positioned within stream systems to oxidize ferrous iron and trap ochre to prevent, or at least diminish, ochre pollution downstream the lake. Although the term "ochre lake" is used commonly, the difference between lake and pond should be explained here. Whether the different ochre abatement measures are ponds, lakes or plants might be also a personal point of view. Lakes are in average deeper than five to seven metres. The pelagial (free water zone) is subdivided into different horizontal thermal layers. Macrophytes occur only within the shallow parts of the lake. Ponds depth (Weiher bzw. Teich) is shallower than five to seven metres. The ground can be overgrown completely by macrophytes. A stabile thermal layering is not possible (POTT & REMY 2000). The ochre abatement plants in Jutland have a maximum depth of two metres. Although it would be biological correct to say plant or pond the term "ochre lake" will be still used in the following chapters.

3.2.2 Volume and retention time

Following chapters are based on information from the Countys Ringkøbing (including the Municipalities Herning and Holstebro) and the SJA. Different factors need to be considered during the installation and maintenance of ochre lakes. The retention time is said to be one of the most important ones. The bigger the volume, the longer the retention time and the better the cleaning efficiency. But the relation between cleaning efficiency and retention time is an exponential not a linear one (HOLSTEBRO MUNICIPALITY, KOFOED personal notification 2005). The recommended retention time is at least eight hours (CHRISTENSEN 1992). The calculation of the volume is time based. Mostly the theoretical median maximum is used to calculate a volume that is sufficient to guarantee a retention time of at least eight hours. Within Holstebro Municipality a retention time of ten to 12 hours is calculated. Lakes have to be emptied less often. The volume is also influenced by the position within the stream system.

3.2.3 Position within the stream system

A lot of measurements are necessary to identify extent and position of the ochre sources. The part upstream the lake is often lost as living space. The higher the lake is positioned within the stream, the lower is the discharge. Less volume is needed what makes the project cheaper. Small lakes with small surface area are less liable to wind disturbance (OKKERGRUPPEN 1999). Another advantage is that a longer stretch is improved as living space for organisms (RINGKØBING COUNTY, JENSEN personal notification 2005). Ochre lakes are often installed to support restoration projects. The installation should also include structural improvements down- and upstream the lakes. If the concentration is still exceeding the threshold limit at the outlet, it might be diluted downstream by another tributary (HERNING MUNICIPALITY, BRANDT personal notification 2005).

The decision about the location and shape is also influenced by landowners or the public. In another case a stream with good chemical water quality was led around the ochre lake to dilute the concentrations downstream the ochre lake. Ochre lakes should ideally enhance landscape amenity and benefit recreation. Ochre lakes can also be installed as compensatory measures (HERNING MUNICIPALITY, BRANDT personal notification 2005).

3.2.4 Maintenance and disposal

In Herning Municipality lakes are cleaned all five to ten years. The ochre sludge is dumped at adjacent fields. Where ochre sludge is deposited vegetation can establish only hardly because of the low pH. Yllebjerg plant at Holstebro Municipality was emptied in spring 2005. During the excavation of ochre the inlet of the lake was closed and water could still flow through an old pipe that was left after the opening of the stream. The ochre is stored for a few days at the adjacent meadow to loose water and volume. Finally it is stored at a **landfill site** (figure 6). The land is principally bought from the landowner and taken out of production (RINGKØBING COUNTY, JENSEN written notification 2005).

Nickel and cadmium occur in critical amounts in the ochre. Because it is strongly bound to the ochre, the possibility of leaching out is little (RINGKØBING COUNTY, JENSEN personal notification 2005). Unnatural concentrations of cadmium and nickel are caused mainly by atmospheric depositions of the metal industries.[9] Cadmium is also released by phosphate fertilizers. Sediments of the ochre lakes in the SJA showed also high concentrations of nickel and cadmium.

---

[9]URL: http://www.waterquality.de/trinkwasser/TWBECHEX.HTM [25.07.05]

Figure 6: After the removal of ochre it is stored for a few days aside the lake. The loss of water reduces the volume. Finally the ochre is stored at a landfill site close to Hvidmose Ochre Abatement Plant, Ringkøbing County. (Picture: H. Prange, 05/2005)

3.2.5 Vegetation within the ochre lakes

Vegetation is said to have a bigger influence on the cleaning efficiency when the pH is low. Vegetation raises the pH and provides living space for microorganisms (catalytic surface). In the SJA pH is naturally high and already oxidized iron needs to be sedimented. Diversity and vegetation cover influence the oxidation efficiency of a lake. In autumn 2004 Herning Municipality investigated 22 ochre abatement plants for iron and chemical parameters as well as the plant cover (Hult-Sernanderske-Method) and diversity (Shannon-Weaver-Index). A positive correlation was found between the coverage within the shallow parts of BGGB-plants (figure 7) and iron concentration as well as the cleaning efficiency (THORDAL-CHRISTENSEN 2004).

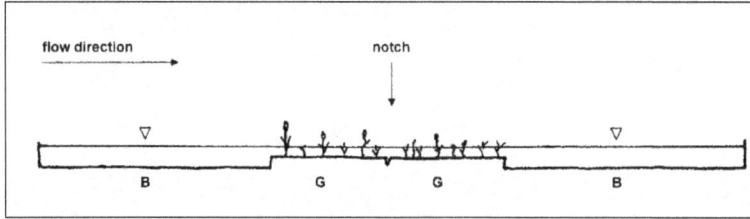

Figure 7: Cross-section in longitudinal direction of a BGGB plant (B = bundfældningsdel = sedimentation compartment; G = grødedel = compartment with weed growth). Notches are constructed within the shallow parts to distribute the current and prevent development of channels within the vegetation.

The development of vegetation is favoured in different ways. In Ringkøbing County new plants are established by storing the top soil layer and incorporate common meadow grass seeds with a milling machine. Finally this is applied to the shallow parts within the lake. In a worse case the development of the grasses can take up to three years, than the lake will be flooded and grasses will be replaced successively by water or wetland plants. Ochre lakes are emptied all four to five years. This means to reset the clock and plants have to establish again (RINGKØBING COUNTY, JENSEN personal notification 2005). In Herning Municipality ten centimetres of top soil is applied on the slopes and shallow parts of the lake without any additional seeds. The lakes are directly flooded. The decay of organic material would cause altered chemical conditions that could be disadvantageous to the cleaning efficiency. Water plants will establish themselves (HERNING MUNICIPALITY, BRANDT personal notification 2005). Channels within the vegetation cause a shorter retention time and less catalytic surface.

3.2.6 Attempts to improve cleaning efficiency

Cleaning efficiency is in general lower during winter season because of higher iron loads and discharges as well as less vegetation. Different attempts aimed to improve the cleaning efficiency. Most successful is raising the pH. During an experiment in spring 2000 the pH was raised from naturally 6.2 – 5.5 to a value of 8.5 by application of NaOH (200g/m³/day). The cleaning efficiency was nearly 100 % than (RINGKØBING AMT 2000). Also in the Environmental Report No. 192 (CHRISTENSEN 1992) efficiencies of 98 % for iron dissolved and 90 % for iron total were mentioned if hydrate lime was applied. The basins of this experiment were 25 cm deep while the retention time was eight hours. The pH > 6 and iron concentration of 15 – 20 mg/l were measured. Other attempts during spring 2000 tried to enhance the catalytic surface. Old Christmas trees were put into the ochre abatement plant at Hvidmose. The cleaning efficiency was increased slightly, but the pH decreased. Also plastic materials were applied to enlarge the surface. "Incamate" was used once at Hvidmose and "Bioblox" were used also once at Ginderskov ochre abatement plant (RINGKØBING COUNTY, JENSEN written notification 2005). Both materials were clogging fast. Because of a big maintenance effort and materials costs theses measures are too expensive. It was also concluded that the enhancement of surface is useful if the pH is raised to > 6. The different parameters that influence the construction of ochre lakes are summarized in table 8.

Table 8: Parameters that should be considered during the installation of an ochre lake

Parameter	Influences on efficiency
Volume	- Big volumes cause longer retention times
High position in the stream system	- less volume is needed (cheaper) - smaller surface area that is exposed to wind - a longer improved stream stretch and free passage
Maintenance and disposal	- different practises - ochre removal all five to ten years - heavy metals occur in unusual high concentrations
High diversity and regular cover of vegetation	- good for chemical oxidation - enlarges catalytic surface - accelerates sedimentation - can reduce retention time extremely - input of organic matter
Improvements	- prevent channel development - enlarge catalytic surface

- Ein größeres Volumen führt zu einer längeren Aufenthaltszeit und damit zu verbesserten Reinigungsleistungen (wenigstens bis zu einem bestimmten Grad).
- Je höher ein Ockersee im Gewässersystem platziert ist, umso weniger Volumen wird benötigt um die gleiche Aufenthaltszeit zu erreichen. Die verbesserte Gewässerstrecke und die Durchgängigkeit werden damit ebenfalls länger. Die Positionierung wird allerdings auch von Kooperationsbereitschaft und zur Verfügung stehendem Land bestimmt.
- Die Unterhaltungspraxis der Seen ist unterschiedlich. Räumungen erfolgen unter Umständen alle fünf bis zehn Jahre.
- Schwermetalle wie Nickel und Cadmium treten im verstärkten Maße im Ockerschlamm auf.
- Eine gleichmäßig verteilte und stark entwickelte Vegetation mit hoher Diversität erhöht die Reinigungsleistung der Ockerseen.
- Versuche die katalytische Oberfläche künstlich zu erhöhen erwiesen sich als zu arbeitsaufwendig und teuer.

3.3 Examples for different types of lakes

This chapter should provide an overview about different types of lakes with their advantages, disadvantages and costs. Lakes are always consistent of at least one deep part (sedimentation basin) and a shallow part. Sand traps are also usual within the new lakes. **Hvidmose at Ringkøbing County** is an ochre abatement plant with two parallel basins of 20 m width. It can be excavated from both sides. The stream was dammed up in a way that makes free passage impossible, but bad water quality upstream the lake would prevent passage of fish or benthic invertebrates (HOLSTEBRO MUNICIPIALTY, KOFOED personal notification 2005). The initial recommended depth was 25 cm. This is too shallow and basins are clogged fast. The shallow parts should be at least 0.5 m deep. Vegetation can cause shorter retention times because of channel development (figure 8) and prevents equal water distribution within the basins.

The Hvirlå Lake is the oldest one in the SJA (installation 1994). It is consistent of a sand trap, two shallow parts and a central deep part. The second shallow part was supposed for after cleaning. This has not been considered as necessary anymore at the SJA. Following lakes have been constructed with a shallow part followed by a deep part (sand trap > shallow > deep). The Hvirlå Lake was investigated detailed under 3.5.1. Ochre lakes within the Herning Municipality are **BGGB-plants** (see chapter 3.2.5) with two deep parts of 1 m depth and a central shallow part of 0.5 m depth. The lakes have an approximate width of 25 m. They can be excavated from both sides (HERNING MUNICIPALITY, BRANDT personal notification 2005).

Most lakes in Holstebro Municipality and Ringkøbing County are consistent of a **deep part followed by a shallow part**, like e.g. Mørre Bæk (two and one meter deep). Yllebjerk Ochre Lake (also Holstebro Municipality) was dammed up. Meanders following the outlet should diminish a high current caused by the relative steep slope. Alder, willow and birch grow naturally already a few years after the installation. Another example is the four hectare ochre lake at Abilå. With an approximately cleaning efficiency of 80 % it gets out 15 t of iron total per year. High coverage of vegetation (the deep part is dominated by *Potamogeton natans*) and a long retention time of 33 hours cause a good cleaning efficiency.

Figure 8: Channel development within basins of the Ochre Abatement Plant Hvidmose, Ringkøbing County. (Picture: H. Prange, 05/2005)

A dam was built across the river valley for the instalment of **Molsted Brook** ochre lake (also Ringkøbing County). The initial maximum depth was 0.95 m close to the dam. Additionally a 1.25 m deep basin was dig at the inlet as sand trap. A gravel pass was constructed around the dam to enable fish and invertebrates to pass the valley. The costs for installation of the dam are low in comparison to other ochre abatement plants (50.000 €). Compensation was paid to the landowner for the wetter conditions at adjacent grasslands. Another advantage is that the volume increases with higher discharges during winter (RINGKØBING COUNTY, JENSEN personal notification 2005). In comparison to Løgumkloster Bæk Ochre Lake at the SJA, this lake has steep slopes and therefore a lower cover of vegetation. Løgumkloster Bæk Ochre Lake was investigated detailed under 3.5.2.

In total 28 ochre abatement lakes have been installed since 1991 in Ringkøbing County (2.7 Mio €, 50 – 75 % financed by the Danish Forest and Nature Agency). The following project has been a part of the rehabilitation of the **river Timm**. The stream collects acid mine drainage from a brown coal mine. The iron content is 240 mg/l and the pH is 3. The application of 10 % slaked lime milk ($Ca(OH)_2$) increases the cleaning efficiency up to nearly 100 %. After the addition of lime milk the water is led through an aeriater. The huge amounts of ochre sludge are collected by a kind of hover and stored within basins (figure 9). At the outlet of the weed covered basins pH > 8 was measured. These kind of plant is extreme expensive. Only the application of lime milk costs 40.000 € per year. The costs for daily maintenance are relatively low here, because the owner of a fish farm downstream the plant is responsible for daily control visits (RINGKØBING COUNTY 1995 & RINGKØBING COUNTY, JENSEN personal notification 2005).

Figure 9: The ochre abatement plant at the River Timm (Ringkøbing County) purifies water of ferrous iron by the application of lime milk. The lime tank and the aeriater can be seen in the background. Ochre sludge is collected by a kind of hover and stored at a neighboured landfill site. (Picture: H. Prange, 06/2005)

- Allen Anlagentypen ist ein tiefer Sedimentationsbereich (ca. 1,0 m oder tiefer) und ein flacher Bereich (ca. 0,5 m) gemeinsam.
- Das Anheben des pH-Wertes oder Belüftung erfolgt aus Kostengründen nur in extrem sauren Mienenwässern mit Eisengehalten von mehr als 100 mg/l.
- Die Durchgängigkeit an den Seen wird gewährleistet, sofern auf Grund der Verockerung stromaufwärts des Sees überhaupt Wanderung stattfinden würde.

3.4 Definition of efficiency

The **efficiency** means the load of iron that is trapped within the lake. Costs can be calculated in Danish Krones or Euros per kilogramme or tonnes. Efficiency means also in how far the concentrations are reduced and organisms can live within the streams again. Threshold limits defined in the Regional Plan (mentioned in chapter 3.1.2). The WFD demands the "good ecological status" that is defined by the abundance of biological quality parameters that is supported by chemical and hydromorphological factors. Therefore efficiency needs to be defined also in a more holistic and long-term view.

The **biological determination of water quality** was not possible before the implementation of ochre abatement measures in several streams at the SJA because of extreme ochre polluted conditions. Table 9 shows in how far streams have been improved according to biological investigations of water quality. Significant improvements need time and long-term investigations are necessary. It has also to be beard in mind that good structural quality and chemical water quality interact with each other.

Table 9: Length of streams that did not allow a biological determination of water quality because of extreme ochre pollution (SJA a 2002)

Period	Length of streams that could not be determined	Amount
93/94	153 km	10.6 %
94/97	125 km	8.7 %
97/00	105 km	7.3 %
Improved	48 km	3.3 %
Total length	1442 km	100 %

Mostly fish fauna (e-fishing) and benthic invertebrates are investigated in combination with the chemical parameters and iron concentrations. For example in September 2004 nine ochre projects (established between 2000 and 2003 at the tributaries of the stream Storå, Holstebro Municipality) were investigated. According to **benthic organisms** improvements have been detected in seven of nine cases. Evaluation of benthic organisms is difficult, because structural conditions can vary down- and upstream the ochre lakes. In one case good structure and high velocity upstream the inlet favours a high abundance of benthic organisms during summer season. But during winter season the conditions are worse due to ochre depositions (HOLSTEBRO MUNICIPALITY, KOFOED personal notification 2005 and KOFOED 2004).

- Effizienz bedeutet zum einen zurück gehaltenes Eisen in Kilogramm oder Tonnen.
- Zum anderen machen Ockerseen die biologische Gewässergütebestimmung in zuvor stark belasteten Gewässern überhaupt erst möglich.

3.5 Detailed investigations of three ochre lakes in the SJA

3.5.1 Hvirlå Ochre Lake
3.5.1.1 Background information
Hvirlå Ochre Lake was established in autumn 1993. The river Hvirlå had the highest priority at the Ochre Action Plan 1994. The catchments area of the lake is 21 km². The yearly mean discharge is 300 l/s, while the 95 % percentile is 550 l/s. The stream target (Gewässerzielsetzung) from spring to the village Ravsted is **B3** (cyprinoids waters), what demands a threshold limit of 0.5 mg/l during winter season (figure 10). Downstream Ravsted until the river Vide it is even **B1** (spawning and nursery area for salmonoids).

The upper part of the Hvirlå is polluted by ochre during the whole year, while in the lower part ochre pollution occurs most intensive during winter time. The Hvirlå Ochre Lake was placed directly downstream of two **former wetlands**, Nørre Ølev and Kasso that were drained during the 1950-1970's. These are fens and "okker and myrmalm" (Raseneisenerz) occurs within the surrounding **sandy soils.** Because wetlands could be clearly defined as ochre sources this ochre lake was classified as **"point source"** within the Ochre Action Plan 1994.

The Little Hvirlå Ochre Lake was established at nearly the same time. It traps iron loads from a small tributary and is also a point source. Ravsted Winter Ochre Lake is existent since winter 2002. The Hvirlå Ochre Lake was **constructed** with a shallow part (of approximately 0.25 m depths) and a central deeper part (with a depth of 1.25 m) for deposition of ochre. Within the shallow part *Potamogeton x nitens* (Glänzendes Laichkraut) was planted. To prevent development of channels, obstacles were installed at in- and outlet. In autumn 2000 the lake was enlarged into northern direction. Sketches are enclosed in the appendix. The inlet of the lake was deepened to 29.70 m DNN to function as a sand trap (SJA 1993).

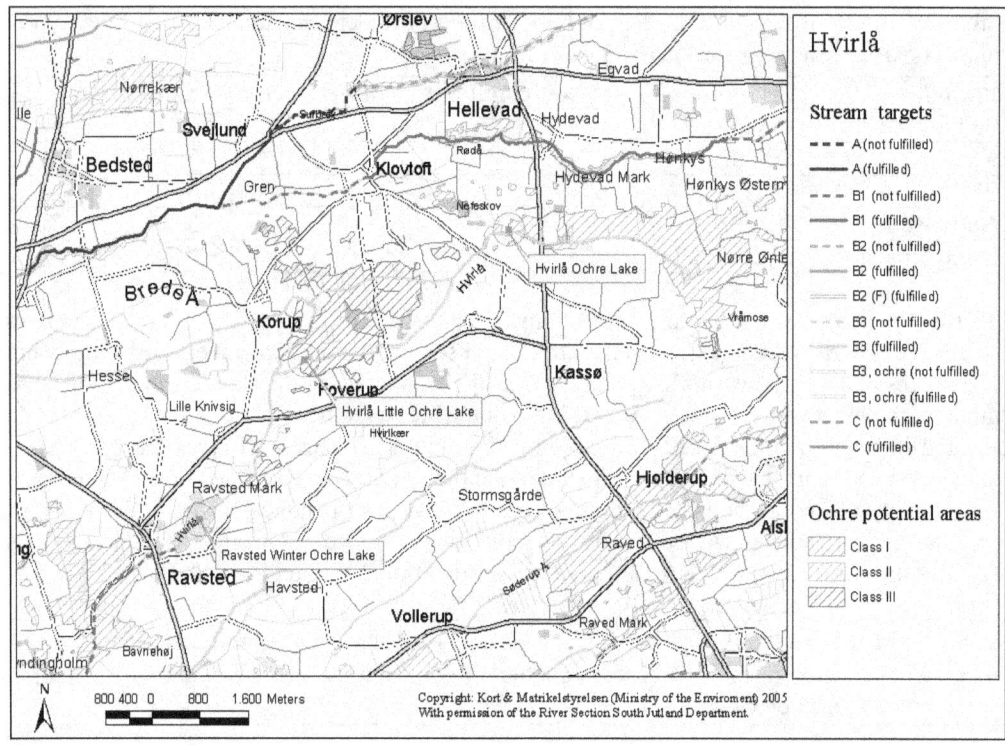

Figure 10: Location of the Hvirlå Ochre Lake, the Little Hvirlå Ochre Lake and Ravsted Winter Ochre Lake. Also shown are stream targets and ochre potential areas. The Hvirlå Ochre Lake is directly located downstream former wetlands.

3.5.1.2 Concentrations of dissolved iron and total iron and the influence of discharges

Data for the Hvirlå Ochre Lake are available from 1993 to now. Dissolved iron reached concentrations of more than 4 mg/l, whereas total iron occasionally reached concentrations of 11 mg/l. Before the enlargement in autumn 2000 the concentration of **dissolved iron** at the outlet was in average **1.8 mg/l** during winter. Since the enlargement the concentration of dissolved iron has been reduced to **0.59 mg/l** during winter seasons (Table A1).

Figure 11 shows the concentrations during winter seasons based on the medians (Table A2b). During the first two years dissolved iron was approximately 4 mg/l and total iron was 7 mg/l. But during the following five years total iron **constantly increased**. Concentrations of dissolved iron increased less intensive and declined in winter 99/00. The **ratio** between dissolved iron and total iron **became bigger**. The concentration of **dissolved iron** at the inlet has been **constantly lower** since winter 99/00, the same year as the enlargement. Total iron concentrations have been more fluctuating. Higher discharges cause higher iron concentrations (Figure 12). The relation between discharge and dissolved iron is stronger than between discharge and total iron.

The **average cleaning efficiency of dissolved iron** during winter seasons was 46 % before and **63 %** after the enhancement (Table A1). The cleaning efficiency of **total iron** was increased from 18 % to **52 %** by the enlargement. Figure 13 shows an improved cleaning efficiency during the first years after the enlargement, especially for dissolved iron. The cleaning efficiency of total iron seems to decrease during the last two winter seasons.

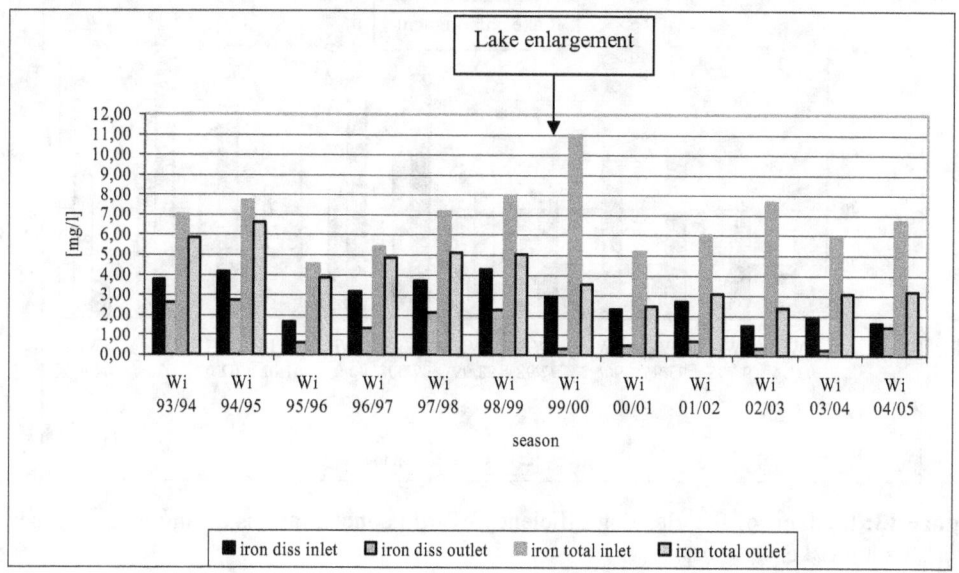

Figure 11: Medians of iron concentrations per winter season at Hvirlå Ochre Lake, 1994–2005.

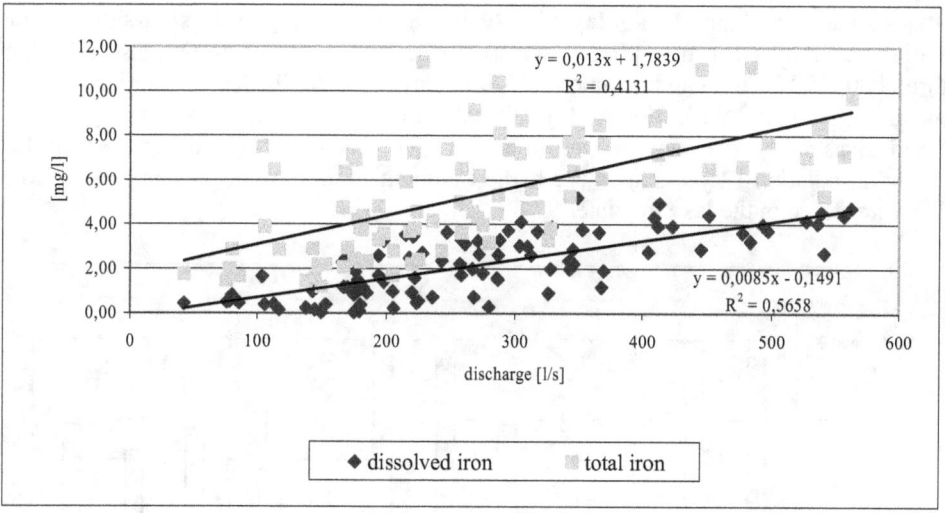

Figure 12: Relation between iron concentrations and discharges at the inlet of the Hvirlå Ochre Lake.

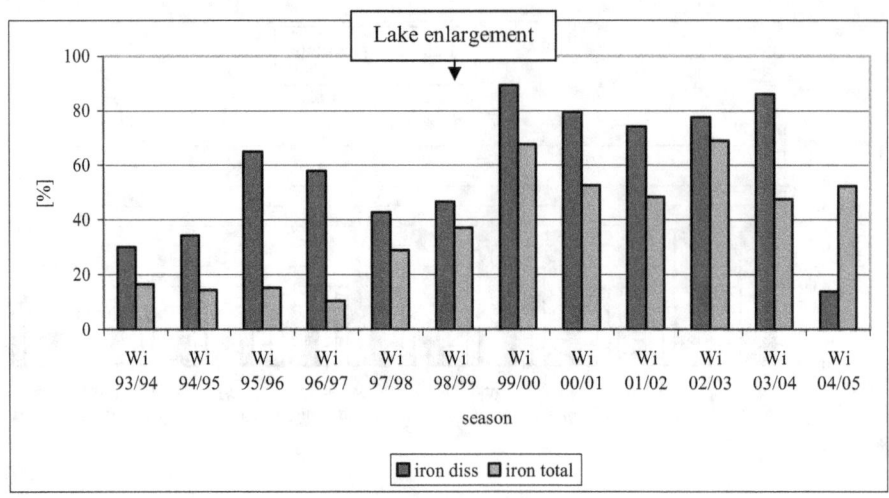

Figure 13: Medians of the cleaning efficiency of iron concentrations during winter season at Hvirlå Ochre Lake.

3.5.1.3 Loads of dissolved iron and total iron at in- and outlet and resulting cleaning efficiencies at Hvirlå Ochre Lake

The load of dissolved iron that reached the lake during winter seasons was in average **109 kg/day** until 2000 and **59 kg/day** after 2000 (Table A4c). Figure 14 shows that the ochre pollution by dissolved iron has been decreased. The load of total iron was also a bit higher before 2000 (222 kg/day) and declined to 176 kg/day since 2000. Ochre pollution reached its **peak in winter 99/00 with 300 kg/day** of total iron, the same year of the enlargement. Table A4d gives an overview of trapped iron loads. The lake has trapped **258 t of total iron** since the installation 12 years ago. The highest amount of total iron (51 t) and dissolved iron (17 t) were stored in the lake in winter 99/00.

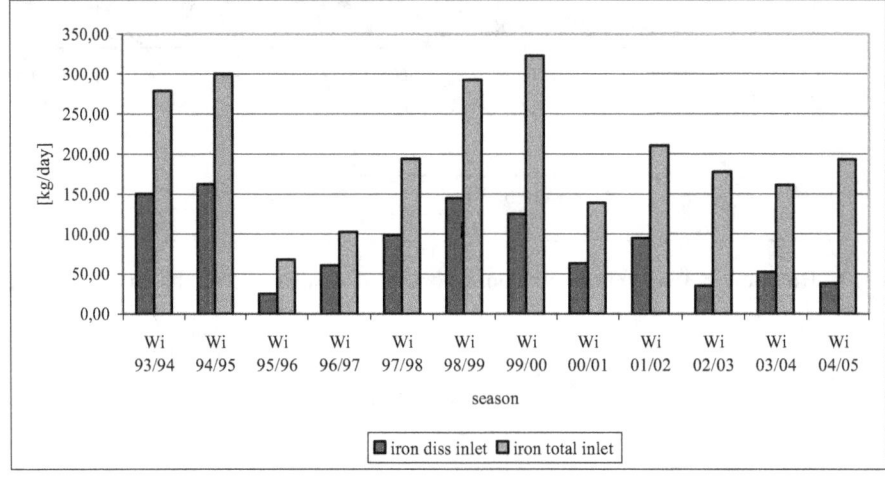

Figure 14: Development of iron loads at the inlet of Hvirlå Ochre Lake.

3.5.1.4 Abundance and distribution of vegetation in the Hvirlå Ochre Lake

Table 10 shows the estimated percentages for vegetation cover. *Potamogeton natans* and *Callitriche* spec. occurred mostly. Three transects were laid across each compartment. Within the **first compartment** *Callitriche* spec. and *Potamogeton natans* were estimated with coverage of 5-25 %. Distribution was regular. Plants were extremely covered by ochre. The old part of **compartment two** was completely overgrown with *Potamogeton natans*. A channel of an approximately width of 10 m occurred within section two, where the new deep part started (figure 15). A strong current occurred there. Within **compartment three** only a few plants were abundant within the old part (*Potamogeton pectinatus*); *Potamogeton natans* appeared with a relative high density within the new part. Using the **Secchi disk** was only possible within the deep parts of compartment two and three. The visibility depth was 1 m in compartment two and 1.7 m in compartment three.

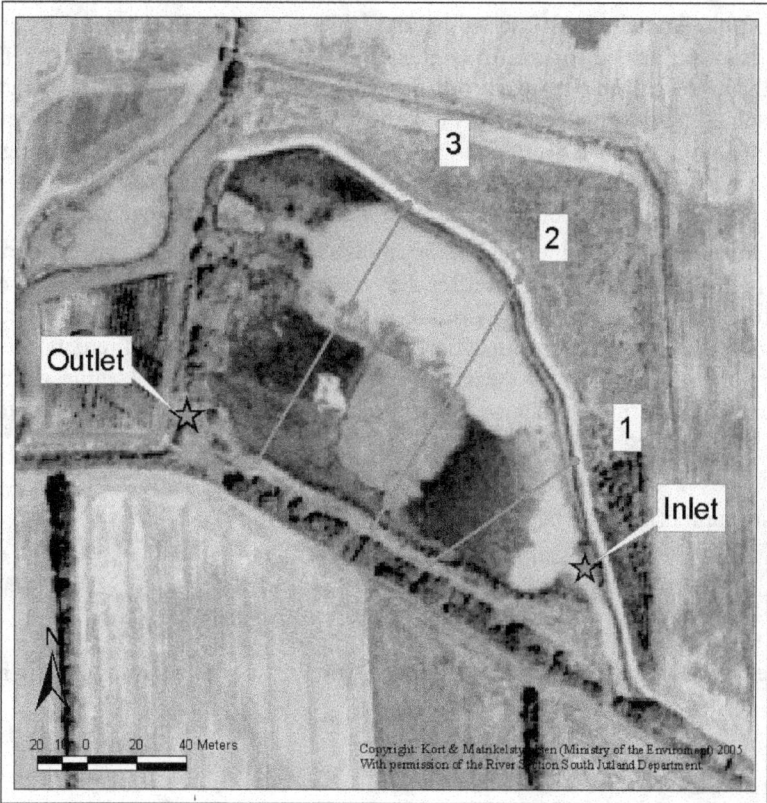

Figure 15: The ortho photograph from 2002 shows the enlargement of the lake in northern direction. Transects 1 to 3 were laid across each compartment to estimate vegetation cover and measure heights of deposition.

Table 10: Estimated plant cover at Hvirlå Ochre Lake

Species	*Potamogeton natans*	*Callitriche* spec.	Green algae	*Potamogenton pectinatus*
Sand trap	5-25 %	5-25 %	0 %	0 %
Compartment 1	5-25 %	50-75 %	> 0-5 %	0 %
Compartment 2	75-95 %	0 %	0 %	0 %
Compartment 3	5-25 %	5-25 %	> 0-5 %	5-25 %

3.5.1.5 Deposition of ochre, retention time and surface-volume-ratio at the Hvirlå Ochre Lake

Table A5 (calculation height of water table) and table A6 to A7 show measured data on which the three cross sections in figure 16 are based. Comparing the initial planned height in m DNN with the recent height **nearly no deposits** occur within the first compartment. The **bottom** is sometimes even **deeper**. Within the second compartment which is partly overgrown by vegetation, deposits were mainly detected **at the edges** (relatively strong deposition appeared at the left side). The old part seems to collect some deposits while the new part is even deeper. Within compartment three nearly no deposits could be detected.

Using 5250 m³ as the old volume and 14.750 m³ as the new volume, the **theoretical retention time** was 7 hours before and 16 hours after the enlargement (Table A9). The theoretical retention time would have been **more as double as long** because of the enlargement. Figure 17 shows that relation between cleaning efficiencies of dissolved iron and retention times is still positive after the enlargement of the Hvirlå Ochre Lake and high cleaning efficiencies are more constant. Considering the theoretical volume and the surface following **surface-volume-ratio** could be calculated as following:

1994 (lake instalment): 1.2 ha = 12.000 m² and 5.250 m³ > 2.3 : 1

2000 (lake enlargement): 1.7 ha = 17.000 m² and 14.750 m³ > 1.15 : 1

The **volume** was **increased** approximately **three times**. The surface-volume-ratio has become smaller (reduced to the half of the initial ratio).

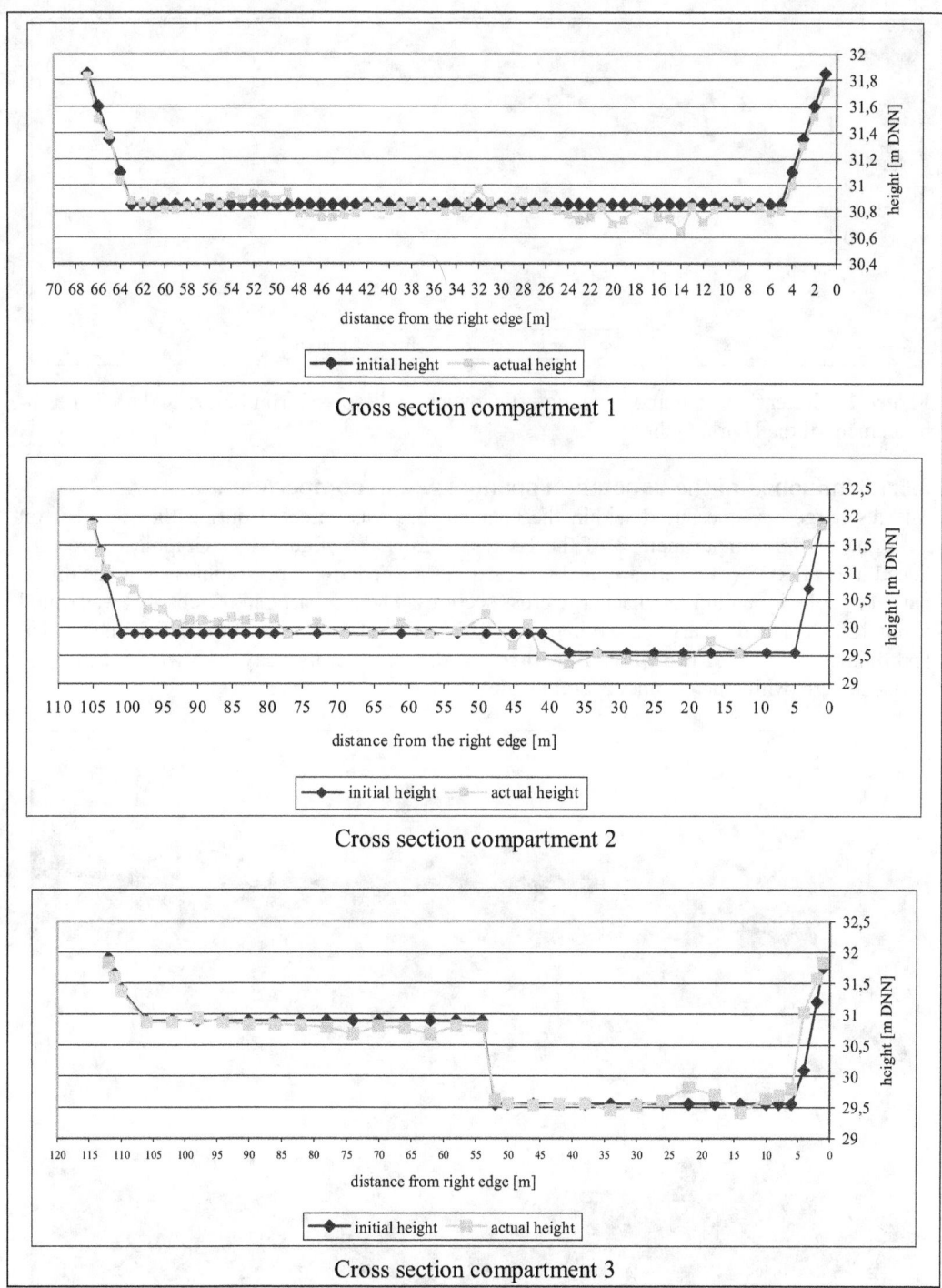

Figure 16: Cross sections of compartment 1, 2 and 3 at the Hvirlå Ochre Lake.

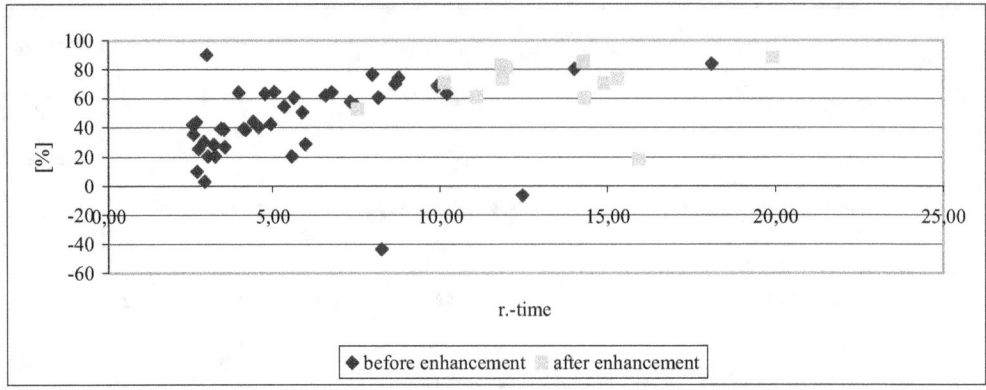

Figure 17: Retention time and cleaning efficiencies of dissolved iron before and after the enhancement of the Hvirlå Ochre Lake.

3.5.1.6 Influence of the vegetation channel within compartment 2

The discharge was measured within the channel that was detected during the site visit on 24th June within compartment 2 of the Hvirlå Ochre Lake (figure 18). Graphics were produced in HYDROS (PC software) and included in the appendix. The graphics show the distribution of velocity within the discharge cross sections. The discharge in the channel was round about 150 l/s. The discharge in the inlet was 210 l/s and at the outlet 190 l/s. This means that approximately **75 % of the water flew through the channel** that day. The width of the lake is 104 m here, while the channel is 10 m wide.

Figure 18: The view from the northern side shows the channel within dense stands of *Potamogenton natans* in compartment 2 of the Hvirlå Ochre Lake. It reduced the total width of 104 m to approximately 10 m. (Picture: H. Prange, 06/2005)

3.5.1.7 Alterations of chemical parameters in the lake

The oxygen content has been in average 7.6 mg/l at the inlet and 8.7 mg/l at the outlet (Table A10). The oxygen concentration was never lower than 4 mg/l. The lake is raising the oxygen concentration during summer as well as winter. Figure 19 shows the medians of oxygen concentration during winter season. A **decrease of oxygen** concentration occurred from winter **95/96 to winter 98/99** at in- and outlet. This was at the same time, when ochre pollution increased due to an altered ratio between ochre and dissolved iron upstream the lake (see figure 11).

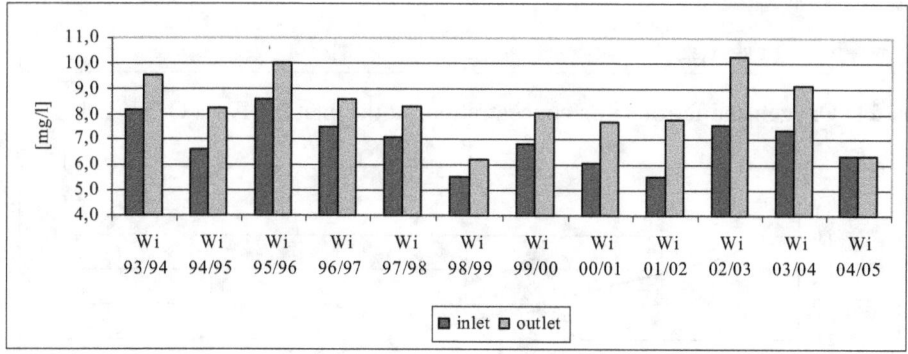

Figure 19: Oxygen concentration at in- and outlet at Hvirlå Ochre Lake.

Also **pH at the inlet decreased** at the same time as oxygen concentrations (figure 20). The pH was **mostly increased** less than 0.1 unit by the lake (Table A11). No correlation between pH and dissolved iron at the inlet could be detected. **Alkalinity** was mostly **decreased** with an average of 0.07 mmol/l (Table A12). The **temperature** has been increased in average 1.2 °C during summer and 0.4 °C during winter season (Table A13). A middle weak negative correlation could be proved for concentrations of ferrous iron and temperature as well as oxygen concentration (figure 21 and 22).

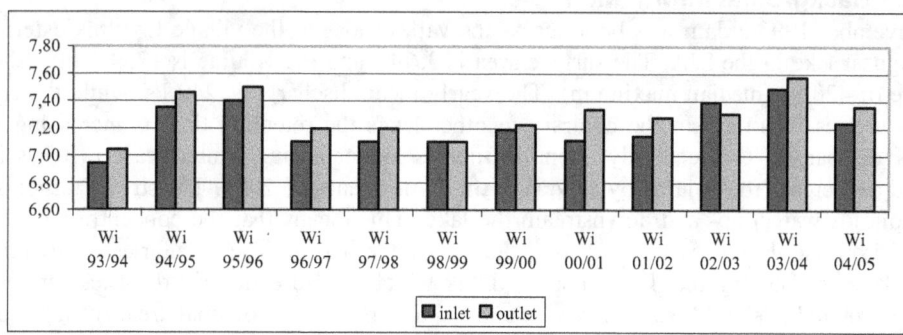

Figure 20: pH at in- and outlet of Hvirlå Ochre Lake.

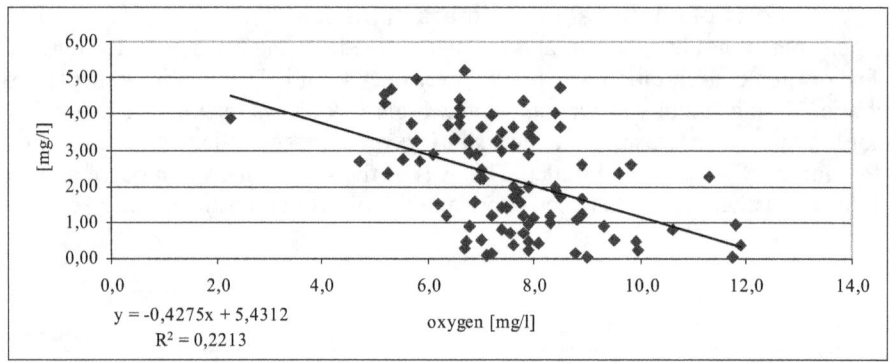

Figure 21: Oxygen and ferrous iron concentrations at the inlet of Hvirlå Ochre Lake.

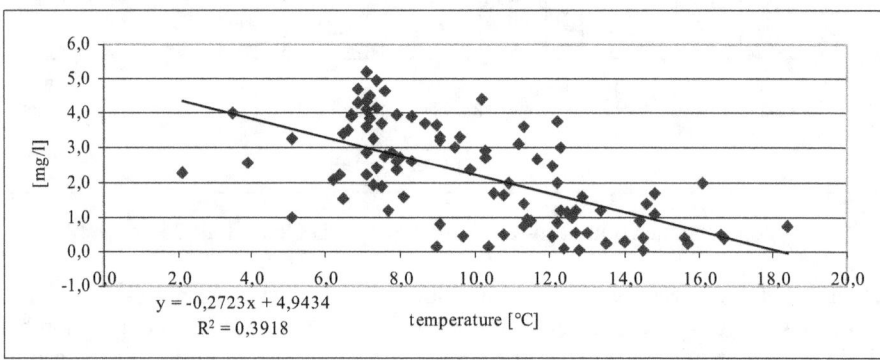

Figure 22: Temperature and dissolved iron concentrations at the inlet of Hvirlå Ochre Lake.

3.5.2 Løgumkloster Bæk Ochre Lake (dam across a valley)
3.5.2.1 Background information

In November 1995 a dam was built across the valley close to the village Løgumkloster. It is the biggest lake in the SJA. The surface area is 3.6 ha and the volume is 12.600 m³ at a discharge of 420 l/s (median maximum). The yearly mean discharge is 240 l/s, while the catchments area is 28.4 km². In the contrary to other lakes the retention time is increasing with higher discharges. Consequently retention time is longer during winter season. It has been planed to increase the volume by **elevating the dam** in autumn 2005. The stream target is **B2** (salmonoids water) up- and downstream the lake. This means that the concentration of dissolved iron should not exceed 0.2 mg/l during winter season. The tributaries upstream this ochre lake are B3 targeted. Løgumkloster Baek which is also called Favrby Bæk, drains the former wetland Alslev Mose. This was identified as a red ochre potential area. A higher input of iron is caused by Skibelund Bæk. The **valley** has been identified as **ochre potential** area class I (figure 23).The soil is consistent of fine sand and organic material in the upper layer (0 – 30 cm) and middle corned sand below (30 – 100 cm). This is tertiary sand that is covered by shifting sands (Flugsande). Different layers can contain varying amounts of iron compounds with different composition. The valley itself is consistent of "fresh water turf "and is surrounded by melt water sands (SJA 1995).

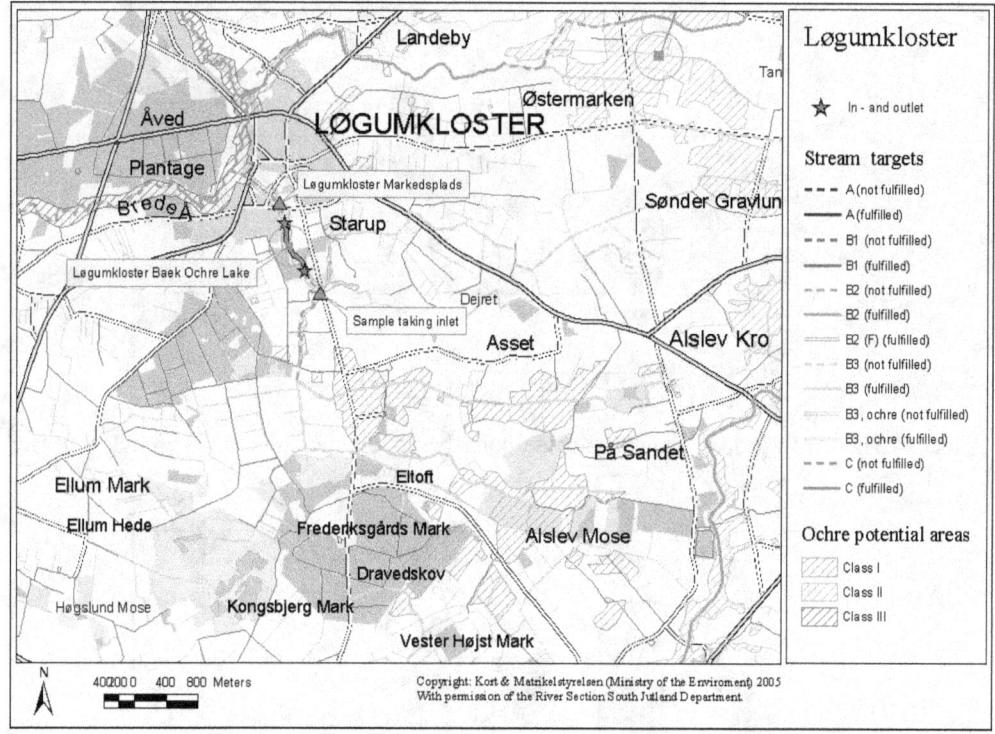

Figure 23: Location of the ochre potential valley and in- and outlet of the Løgumkloster Baek Ochre Lake. Several small tributaries flow into Løgumkloster Baek and therefore the lake has been classified as an ochre lake that traps iron loads from diffuse sources.

3.5.2.2 Concentration of dissolved iron and total iron and influence of discharges

Data of the Løgumkloster Bæk Ochre Lake were only available from November 1995 to December 1999. Two measurements took place in winter '94. Concentration of **dissolved iron** has been in average 1.34 mg/l at the inlet and **0.65 mg/l** at the outlet. The average concentration of dissolved iron at the outlet has been 0.81 mg/l during winter season. **Total iron** has been in average 3.93 mg/l at the inlet and **2.21 mg/l** at the outlet (Table A14).

During winter season total iron can reach concentrations of 5 to 7 mg/l while dissolved iron can reach values of more than 2 mg/l. Medians of total iron at the inlet show an incline during the first four winter seasons (figure 24). The cleaning efficiency decreased extremely during winter 1999 and winter 04/05 (figure 25). Relation between increasing discharges and iron concentrations are positive (figure 26).

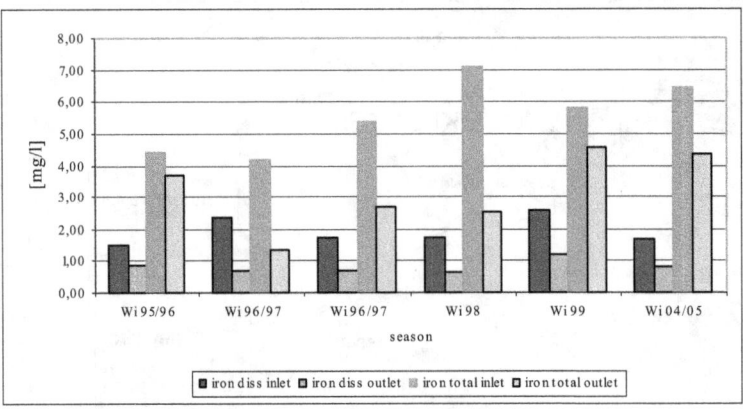

Figure 24: Concentration during winter seasons based on medians at Løgumkloster Bæk Ochre Lake.

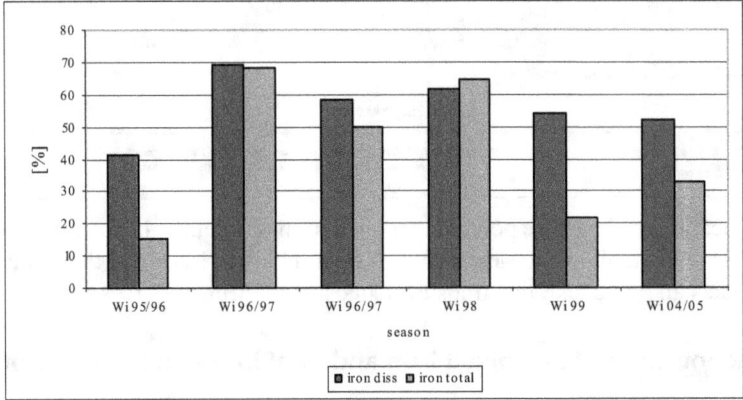

Figure 25: Cleaning efficiency of dissolved iron and total iron at Løgumkloster Bæk Ochre Lake.

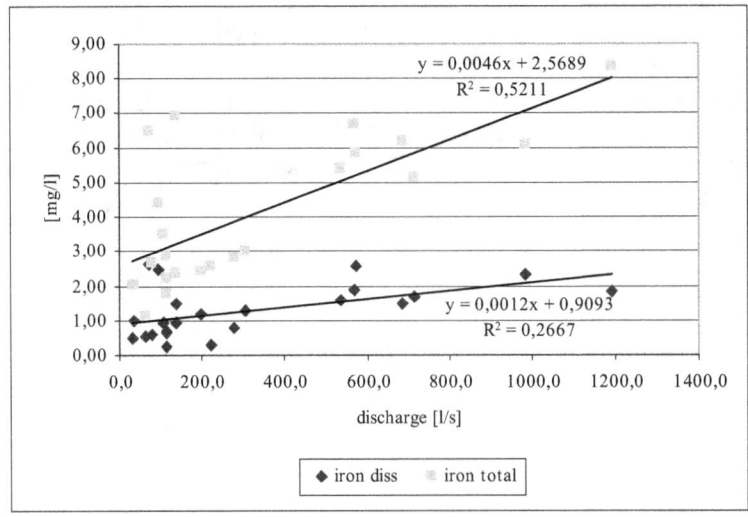

Figure 26: Discharges and iron concentrations at the inlet of Løgumkloster Bæk Ochre Lake.

3.5.2.3 Loads of dissolved iron and total iron at in- and outlet and resulting cleaning efficiencies at Løgumkloster Bæk Ochre Lake

The load of dissolved iron that reached the lake was 98 kg/day and 298 kg/day of total iron during winter season (table A17b and figure 27). The maximum load of total iron occurred in winter 96/97 and was 514 kg/day. Table A17c gives an overview about the amount of iron that has been trapped during the years of measurement. During winter season 96/97 total iron reached its peak with 103.7 t per season. Cleaning efficiency of iron loads was in average 50 % for dissolved iron during winter as well as summer. The cleaning efficiency for total iron is worse during summer (39 %) und better during winter season (53 %).

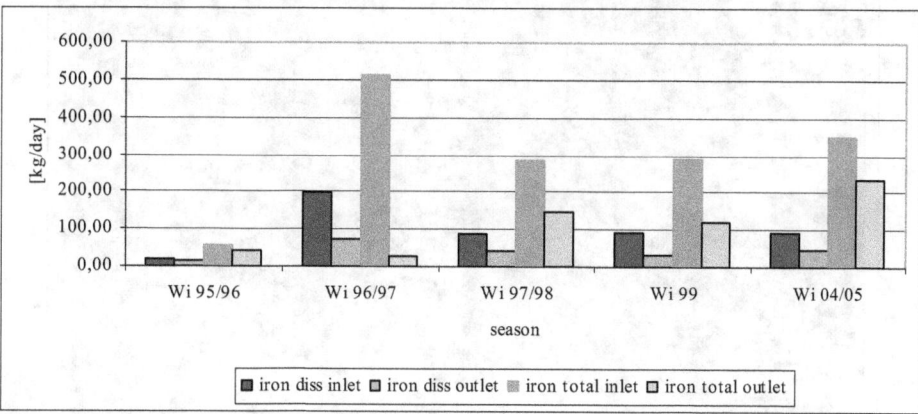

Figure 27: Iron loads in kg/day during winter seasons at Løgumkloster Bæk Ochre Lake (based on medians).

3.5.2.4 Calculation of retention times and surface-volume-ratio of Løgumkloster Bæk Ochre Lake

The inlet of the lake does not contain a sand trap. Huge amounts of sand and ochre have already been deposited here. Investigation of plant cover and diversity in detail was not done here. As it can be seen in figure 28 vegetation is invading the lake extremely from the inlet and the old channel as well as from the shallow slopes. **Vegetation, deposits and channel development** were assumed to decrease the volume. Because the volume and consequently the retention time are said to be the most important factor, it should be investigated in how far the planned enhancement could prolong the retention time. The sketch (appendix) shows the eight compartments, in which the lake was subdivided. Table A18a – A18h and figure 29 show the cross sections that were produced. Table A19a- 19h are the calculations of the volume for every compartment. The volume in total was 13.700 m³ at a height of 10.45 m DNN, 40.157 m³ at a height of 10.75 m³ DNN and 78.671 m³ at 11.00 m DNN (Table A20).

The measured heights of the deposits within the compartments are shown in Table A21. A volume of **11.780 m³** was calculated for the deposits. The thickest layer was measured within compartments 1 to 3 at the upper part of the lake. Finally this volume could be subtracted from the theoretical volume and the retention times could be calculated. **Retention time** has been extremely reduced by the deposits. With high discharges (95 % percentile) the duration of **8 hours at 10.75 m** would be sufficient. An enhancement to **11.00 m DNN** would prolong the retention time up to **~ 20 hours**. The surface-volume-ratio will become smaller if the dam will be elevated (table 11). Extreme changes will take place within compartments 2, 3 and 4. In total the ratio between surface area and volume will be reduced from 3.1:1 to 1.4:1 (table 12).

Figure 28: The ortho photograph from 2002 includes transects within compartments 1 to 8 at Løgumkloster Bæk Ochre Lake. Vegetation is invading the lake especially from the inlet.

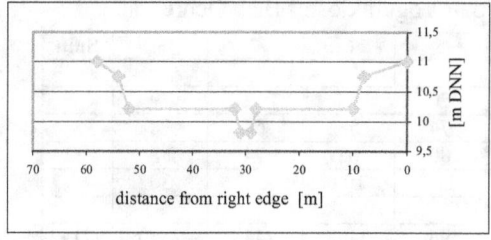
Cross section of compartment 1

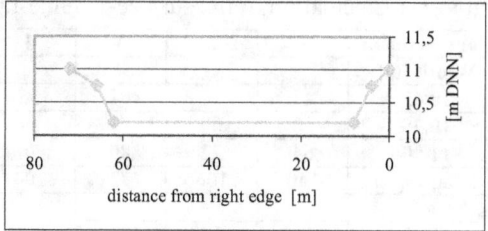
Cross section of compartment 2

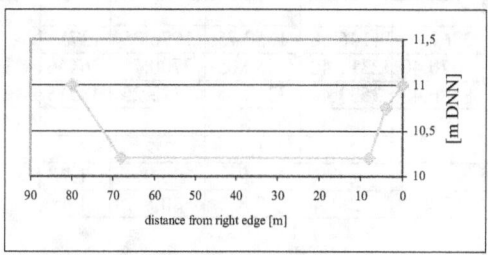
Cross section of compartment 3

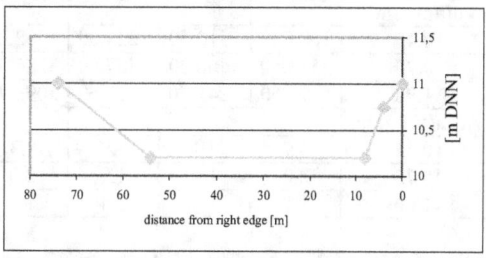
Cross section of compartment 4

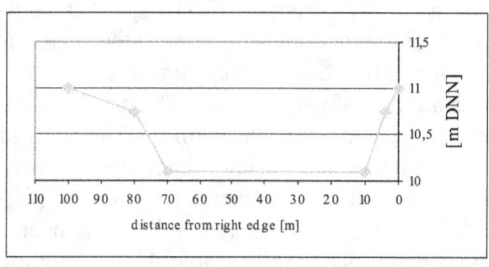
Cross section of compartment 5

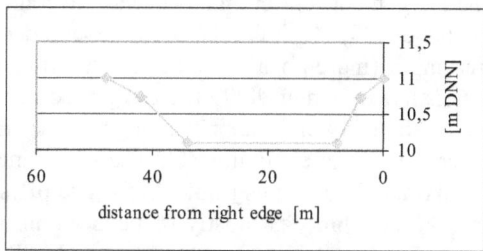
Cross section of compartment 6

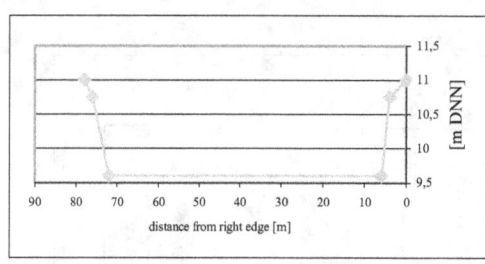
Cross section of compartment 7

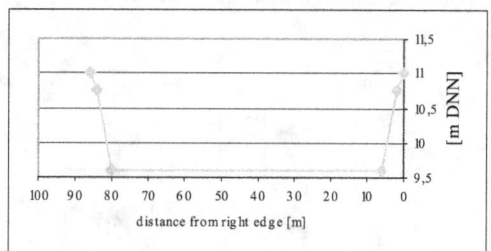
Cross section of compartment 8

Figure 29: Cross sections within Løgumkloster Bæk Ochre Lake.

Table 11: Calculation of retention time and recent volume of Løgumkloster Bæk Ochre Lake

Height [m DNN]	Theor. Volume [m³]	Lost Vol. [m³]	Actual Vol. [m³]	Percentile [%]	Q [l/s]	Theor. R.-time	Actual R.-time
10.45	13710.8	11778	1933	50	325	11.7	1.7
10.45	13710.8	11778	1933	average *	395	9.6	1.4
10.75	40157.10	11778	28380	95	950	11.7	8.3
11.00	78671.90	11778	66894	95	950	23.0	19.6

* This is the average, whereas the 50 %-percentile is the median

Table 12: Calculation of the surface-volume-ratio at Løgumkloster Bæk Ochre Lake

Compartment	1	2	3	4	5	6	7	8	Sum
Width [m]									
10.45	44	58	67	56	70	52	72	79	
10.75	50	60	73	64	74	58	68	78	
11.00	64	72	80	76	80	68	74	88	
Length [m]	140	106	72	104	128	104	34	30	
Area [m^2]									
10.45	6160	6148	4824	5824	8960	5408	2448	2370	42142
10.75	7000	6360	5256	6656	9472	6032	2312	2340	45428
11.00	8960	7632	5760	7904	10240	7072	2516	2640	52724
Volume [m^3]									
10.45	1683.50	1510.50	1224.00	1352.00	2867.20	1146.60	1950.75	1976.25	13711
10.75	3573.50	3466.20	2757.60	3286.40	5670.40	2311.40	2680.05	2700.75	26446
11.00	5393.50	5241.70	4125.60	5093.40	8486.40	3533.40	3317.55	3323.25	38515
Area/Volume									
10.45	3.7	4.1	3.9	4.3	3.1	4.7	1.3	1.2	3.1
10.75	2.0	1.8	1.9	2.0	1.7	2.6	0.9	0.9	1.7
11.00	1.7	1.5	1.4	1.6	1.2	2.0	0.8	0.8	1.4

3.5.2.5 Alterations of chemical parameters within Løgumkloster Bæk Ochre Lake

The oxygen concentration has been in average 7.87 mg/l at the inlet and 7.71 mg/l at the outlet (table A23). As it can be seen the lake decreases the oxygen concentration rather than increasing it (more positive values at the difference) and extreme low concentrations occur occasionally. In winter 98/99 the oxygen content was already low at the inlet. During the other events the oxygen content was high at the inlet. The oxygen concentration has declined in general at the inlet (figure 30). The pH is increased by the lake during winter seasons with approximately 0.1 units (table A25). The pH seems also to decrease upstream in general (figure 31). Alkalinity is mostly reduced within the lake (table A24). Extreme altered temperatures of more than 4 °C occurred occasionally during summer months (table A26). Only the temperature influences the concentration of ferrous iron at the inlet. Concentrations of dissolved iron and temperatures are shown in figure 32.

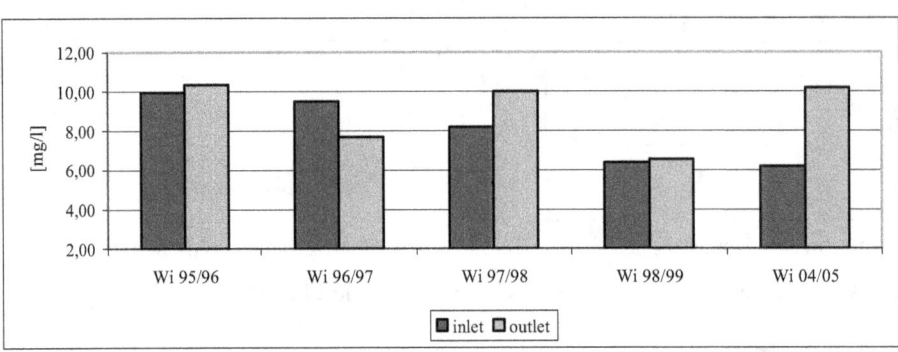

Figure 30: Oxygen concentration at in- and outlet of Løgumkloster Bæk Ochre Lake during winter season.

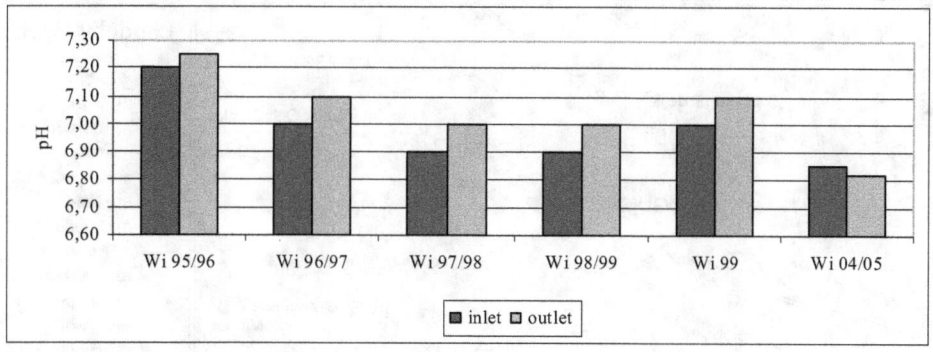

Figure 31: pH at in- and outlet of Løgumkloster Bæk Ochre Lake.

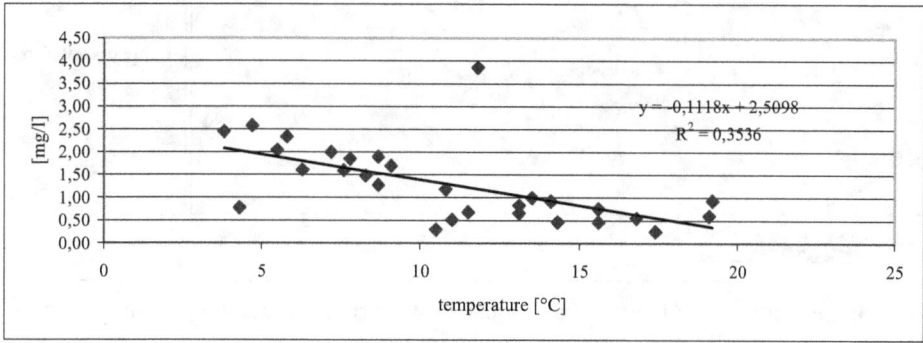

Figure 32: Temperature and dissolved iron concentrations at the inlet of Løgumkloster Bæk Ochre Lake.

3.5.3 Landeby Bæk Ochre Lake
3.5.3.1 Background information
Landeby Bæk Ochre Lake was established in 1999. It consists of a sand trap, a shallow part and a following deep part. The shallow part is 0.4 m deep (1/3 of the lakes area), while the deep part is 1.4 m deep (2/3 of the lakes area). Its surface area is 9.900 m². The volume is approximately 6.804 m³. The yearly **mean discharge is extreme low** (10 l/s). The 95 % percentile is 105 l/s. The **catchments area is 3 km²** in which the soil is mostly sandy (melting sands).

The ochre lake is **located in a former wetland** area that was identified as a green ochre potential area (figure 33). The ochre lake was installed at a small tributary that flows into Landeby Bæk, which is targeted with B2 (salmonoids water) and should not exceed concentrations of 0.2 mg/l of dissolved iron during winter season. The lakes surface area is extremely exposed to wind (SJA 1999). The lake was chosen for investigations because of its **shape** and **extreme good cleaning efficiencies**.

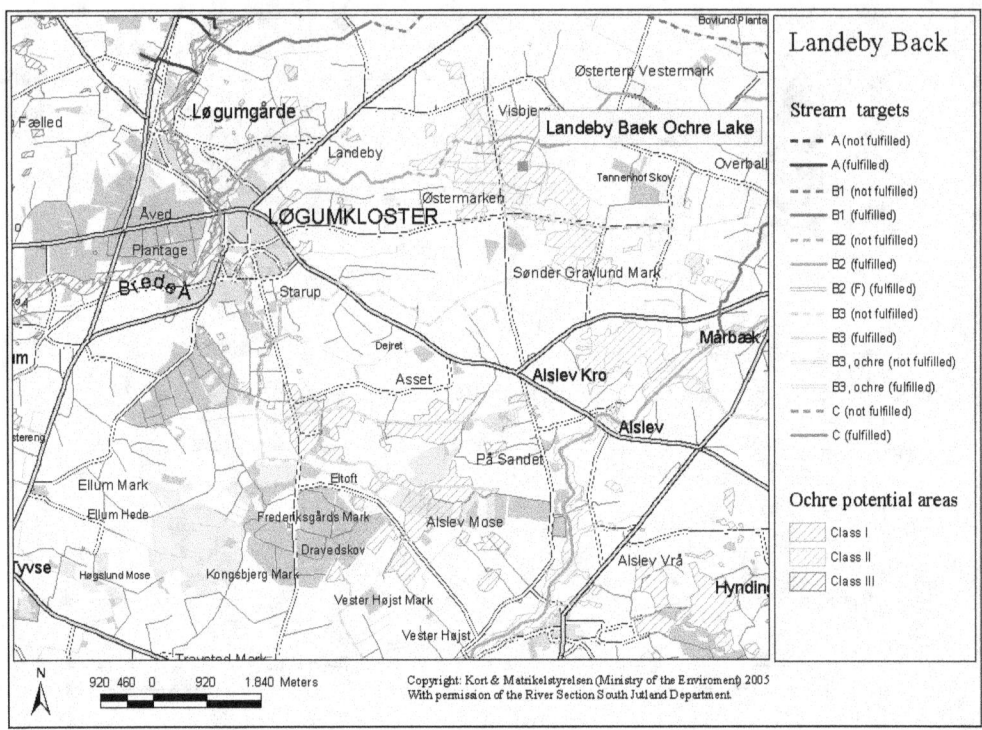

Figure 33: Landeby Bæk Ochre Lake is located at a private watercourse that has no stream target. It drains a class II ochre potential area (point source).

3.5.3.2 Concentrations of dissolved iron and total iron and the influence of discharges

For Landeby Bæk Ochre Lake data were available from 2001 to 2004. The concentration of **dissolved iron** has been reduced **from 2.01 to 0.12 mg/l** during winter seasons. The concentration of total iron was 4.39 mg/l at the inlet and has been decreased to 2.11 mg/l during winter season (figure 34 and table A27). Concentration of dissolved iron is in general extremely low at the outlet (independent on the season).

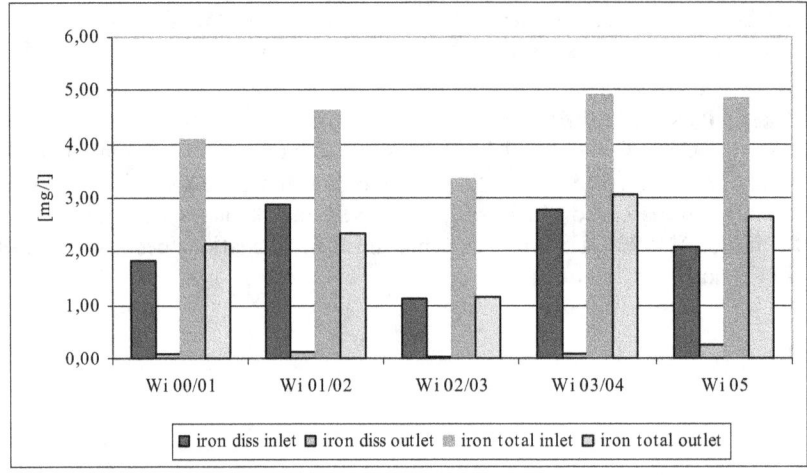

Figure 34: Concentration of dissolved iron and total iron at Landeby Bæk Ochre Lake.

Figure 35 shows that the **cleaning efficiency of dissolved iron** is always close to **100 %**. In comparison to that the reduction effect for total iron drops down regularly during winter season to approximately 50 %. On the contrary to the investigation of the other two lakes all available data were used here because of the small amount of measurements. Discharges show a positive relation to dissolved and total iron at the inlet (figure 36). Concentrations of dissolved iron are also dependent on daily precipitation (figure 37).

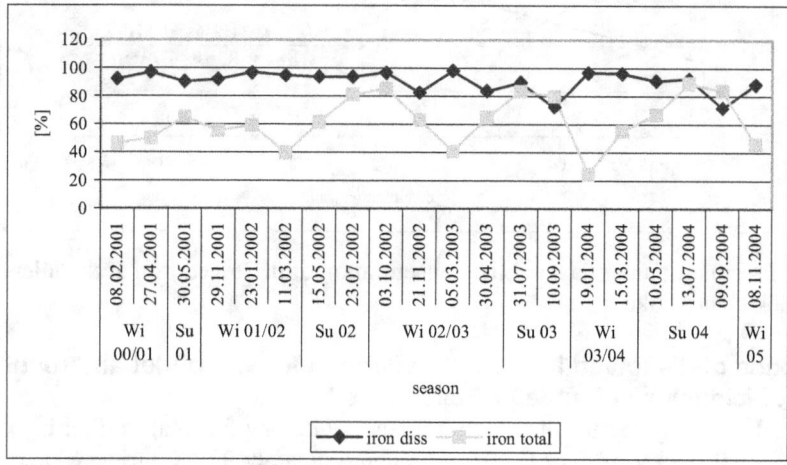

Figure 35: Cleaning efficiency of iron concentrations at Landeby Bæk Ochre Lake (based on all data).

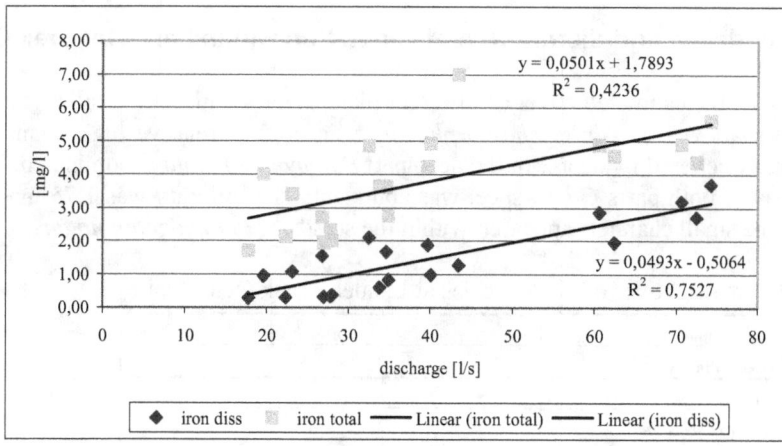

Figure 36: Discharges and iron concentrations at the inlet of Landeby Bæk Ochre Lake.

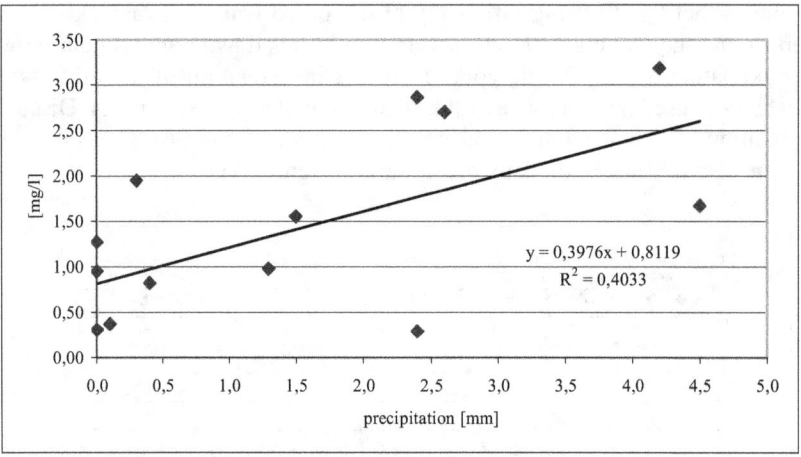

Figure 37: Precipitation per day and the concentration of dissolved iron at the inlet of Landeby Bæk Ochre Lake.

3.5.3.3 Loads of dissolved iron and total iron at in- and outlet and resulting cleaning efficiencies at Landeby Bæk Ochre Lake

The iron load is 9.5 kg/day for dissolved iron and 19 kg/day for total iron during winter season. Cleaning efficiency of dissolved iron is approximately 90 % during winter as well as summer (Table A31b). The amount of total iron that has been trapped in the lake during the five years is round about 9200 kg, while the amount of dissolved iron is higher with 10.400 kg (Table A31c).

3.5.3.4 Abundance and distribution of vegetation within Landeby Bæk Ochre Lake

Table 13 shows the estimated cover of different plant species within the shallow and the deep part. In the shallow part *Callitriche corphocarpa* occured in relative high abundances and were extremely covered by ochre. In the deep part *Potamogeton natans* dominated. Within the shallow areas of both parts *Chara* spec. was abundant. The turbidity was 1.75 m, as deep as the lake. Some small channels appeared within the stands of *Potamogeton natans*.

Table 13: Estimated cover of plant species at Landeby Bæk Ochre Lake

Shallow compartment	Cover
Species Latin Name	[%]
Typha spec.	0-5
Callitriche cophocarpa	25-50
Potamogenton natans	0-5
Chara globularis	5-25
Potamogeton pectinatus	0-5
Deep compartment	
Species Latin Name	[%]
Typha spec.	0-5
Callitriche cophocarpa	0-5
Potamogenton natans	50-75
Chara globularis	5-25
Potamogeton pectinatus	0-5

3.5.3.5 Calculation of retention times and surface-volume-ratio

Measured height of the water table at the inlet was 0.4 m. Two transects were laid across the shallow and the deep compartment to investigate location and height of actual depositions (figure 38). Table A33a to A33d summarize data on which the cross sections (figure 39) are based. The ochre layer had an approximate thickness of 20 cm in the shallow and 30 cm in the deep part. Deposition is relatively **regular distributed** but occurs more extremely on the edges. The sediment core was taken at transect 1 and showed different layers (figure 40). The average thickness of ochre layers within the deep and shallow compartment was multiplied with the surface area (3300 m² at the shallow part and 6600 m² at the deep part) to get the actual volume of 4000 m³ (Table A34).

The theoretical retention time is 48 hours during winter season. The relation between the cleaning efficiency of total iron and retention times is positive, while the cleaning efficiency of dissolved iron is constantly high and independent of retention times (figure 41). The retention time has been reduced by **nearly 40 %** due to less volume (Table A36). The relation between surface area and volume is 1.46: 1 (9900 m² divided through 6804 m³).

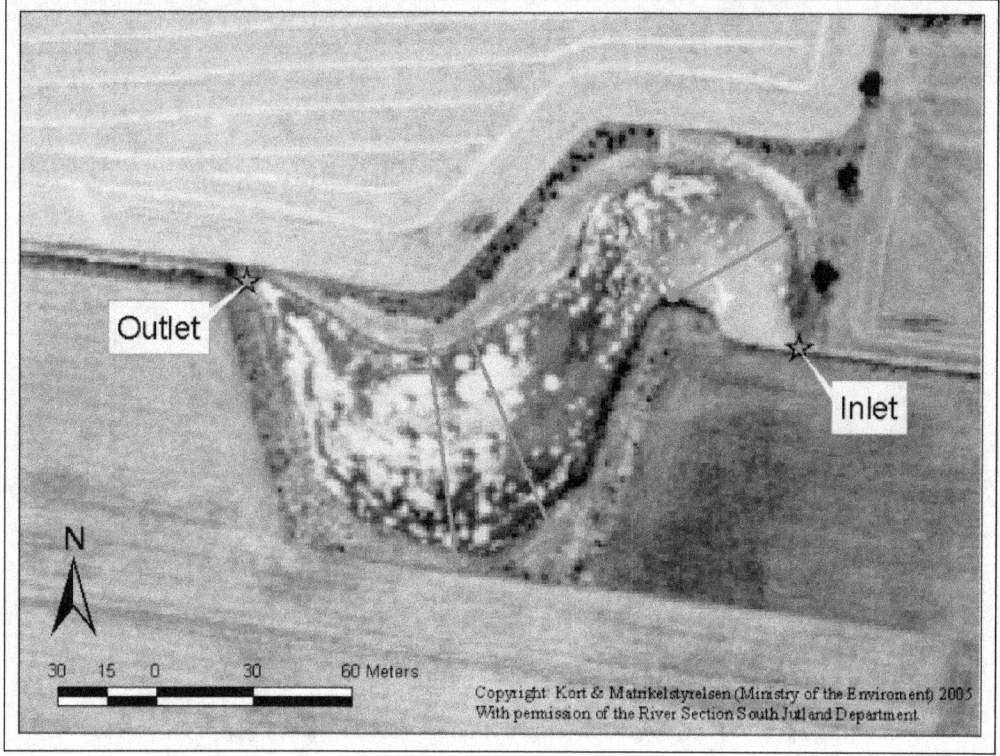

Figure 38: Two transects were laid across the deep and the shallow compartment. The ortho photograph from 2002 shows the ochreous coloured water at the inlet of the lake.

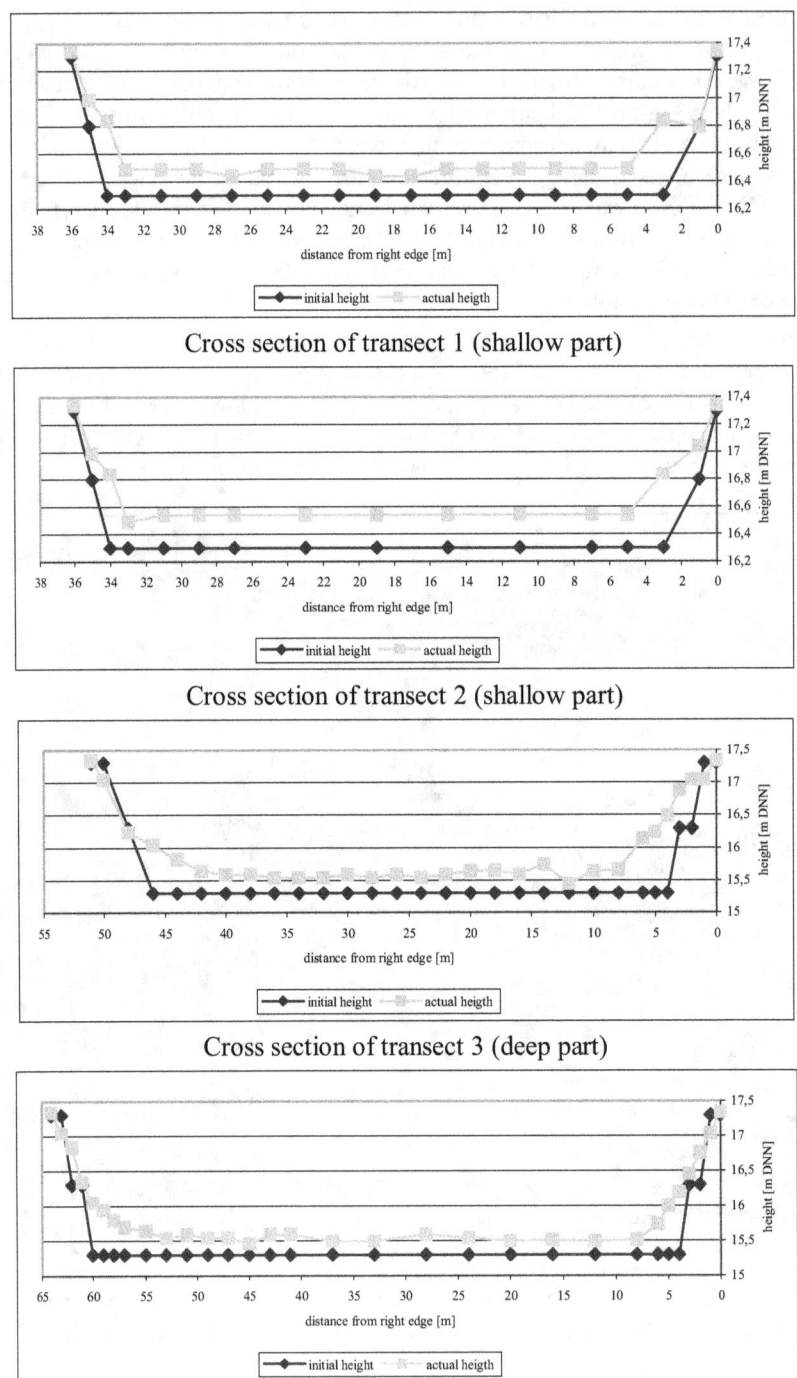

Cross section of transect 1 (shallow part)

Cross section of transect 2 (shallow part)

Cross section of transect 3 (deep part)

Cross section of transect 4 (deep part)

Figure 39: Cross sections 1 to 4 within compartment 1 and 2 of Landeby Bæk Ochre Lake.

Figure 40: Sediment cores consist of a sandy layer covered by organic matter. The ochre lies on the top.(Picture: H. Prange, 06/2005)

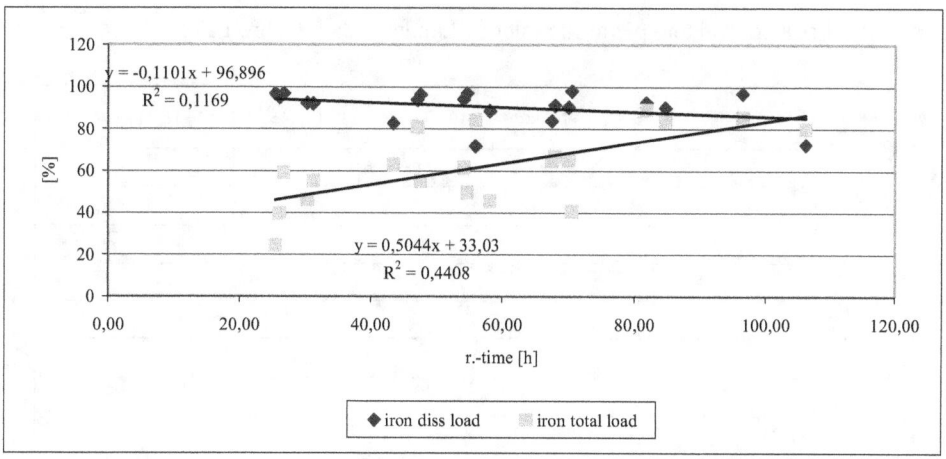

Figure 41: Relation between retention time and cleaning efficiency of iron loads at Landeby Bæk Ochre Lake.

3.5.3.6 Alterations of chemical parameters at Landeby Bæk Ochre Lake

Oxygen concentrations are in average 2.5 mg/l higher at the outlet of the lake (table A37). Also the pH is increased by 0.3 units (table A38). Temperatures increased occasionally by more than 3°C during summer (table A39). A relative strong negative relation could be shown between temperature and concentrations of dissolved iron (figure 42). The relation between pH and dissolved iron at the inlet is positive (figure 43). Additionally the oxygen concentrations at the deepest part were measured during the site visit on 22nd June 2005. The temperature decreased by 4 °C within a depth of 1.4 m. pH did not change. Alterations of oxygen were minimal. No layering occurred (figure 44 and table 40).

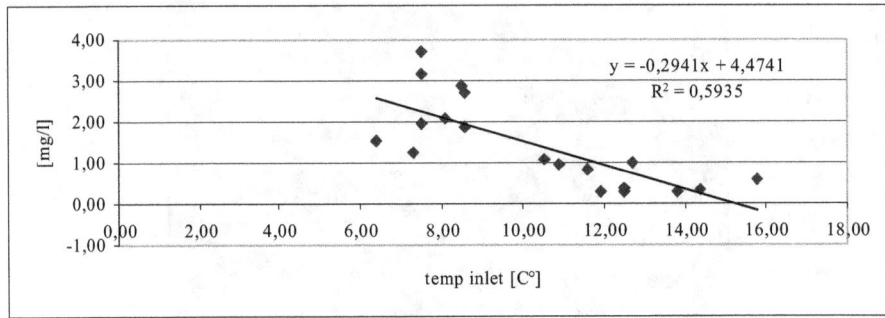

Figure 42: Dissolved iron and temperature at the inlet of Landeby Bæk Ochre Lake.

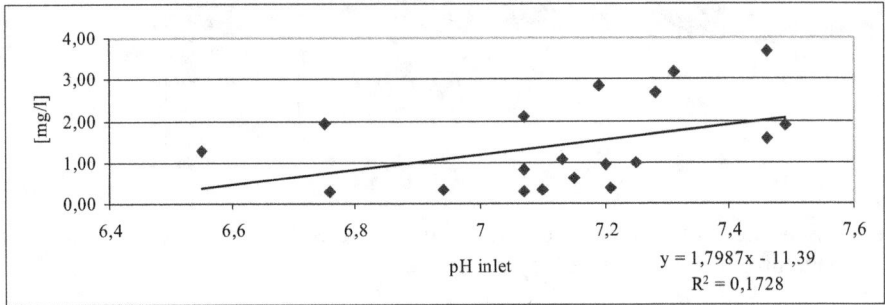

Figure 43: Dissolved iron and pH at the inlet of Landeby Bæk Ochre Lake.

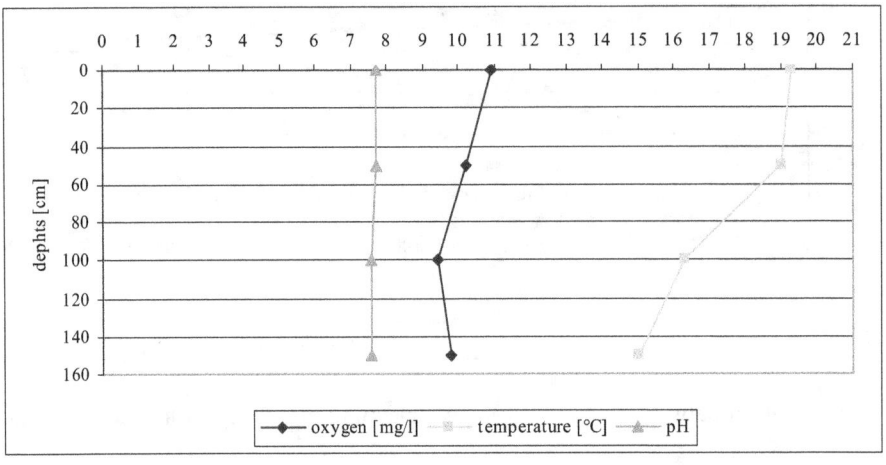

Figure 44: Depth profile at the deepest part of Landeby Bæk Ochre Lake, 22nd June 2005.

3.6 Further possibilities to combat ochre pollution

3.6.1 Winter Ochre Lakes

Winter Lake means that a stream is dammed up during the winter period. This causes flooding of adjacent meadows. Ochre and nutrients are deposited here. The stream is cleaning itself. During summer season meadows are still grazed. Nevertheless water tables can fall during dry summer seasons and pyrite can be oxidized. As an example the **Ravsted Winter Ochre Lake** should be mentioned here. The stream target is B3(F) upstream and even B1 downstream the village Ravsted. As already mentioned under 3.5.1.1 the Ravsted Winter Lake is one of three ochre lakes at the Hvirlå stream. While the other two ochre lakes trap iron loads released from former wetlands in the upper catchment area, ochre pollution further downstream is diffuse (MØLLER & THOMSEN 1992).

Ravsted Winter Ochre Lake was first flooded during winter 2002. The stream is dammed up from **15th October to 15th March** (figure 45). The project area at Ravsted is 28 ha in total. The flooded area varies between four and 18 ha, depending on precipitation. The catchments area is 31.2 km² at the inlet of the project area. Free passage of fish and benthic invertebrates is guaranteed by a gravel by-pass around the dam. Five landowners are involved in the project. The project area includes an §3 area, that was designated as wet meadow (Feuchtwiese). § 3 of the Danish Nature Protection Law means that areas of special nature types bigger than 2.5 ha are not allowed to be altered.

Figure 45: A weir was installed in 2002. The Hvirlå is usually dammed up from 15th October to 15th March. Free passage is provided by a gravel by-pass. (Picture: H. Prange, 03/2005)

Usually landowners do not get **compensation** within § 3 areas. Because the instalment of an ochre lake influences land use practises the landowners received an initial compensation calculated with 50 – 80 % of the recent market value (SJA RIVER SECTION, HANSEN written notification 2005). The lake cost 168.000 €, including 92.800 € compensation payments. 75 % of the costs are financed by the Danish Nature and Forest Agency (SJA b 2002). Natural occurring wet meadows and groundwater influence favoured the instalment of the winter lake. It supports the other two lakes and makes it more likely to reach the stream target of 0.2 mg/l ferrous iron downstream Ravsted.

3.6.2 Combat the origin of ochre pollution – raise water tables

Raising water tables would cause anaerobic conditions again. Pyrite can not be oxidized anymore. A higher water table can be reached by different methods. One is to cut drainage pipes at valley edges. It is also possible to make the **channel cross section wider but shallower** during actual restoration projects as shown in figure 46 (HOLSTEBRO MUNICIPALITY, KOFOED personal notification 2005). Another possibility is to **apply stones** on the channel bottom and raise the channel cross section as well as the water table. Applied measures need to fulfil descriptions in the regulative. If the channel cross section is deeper it is also possible to **leave more weeds** in the channel what reduces the volume and causes a higher water table, too. The possibility of altered maintenance will be described in detail in the following chapters and should always be considered as an additional and supporting measure.

Choosing the right measure to combat ochre pollution should ideally be based on the kind of ochre source (diffuse or point source) and characteristics of the catchments area (pyrite content within soil, hydrogeology etc.). It demands investigations and monitoring of iron concentration and loads in advance as well as control of measures efficiency. Raising water tables to cause anaerobic conditions again is not always possible, because the choice of measure is also dependent on landscape and terrain (clearly demarked valley or flat terrain) and stakeholder interests. Following chapters will describe different projects which have had the aim to diminish ochre pollution during the last 12 years in Jutland.

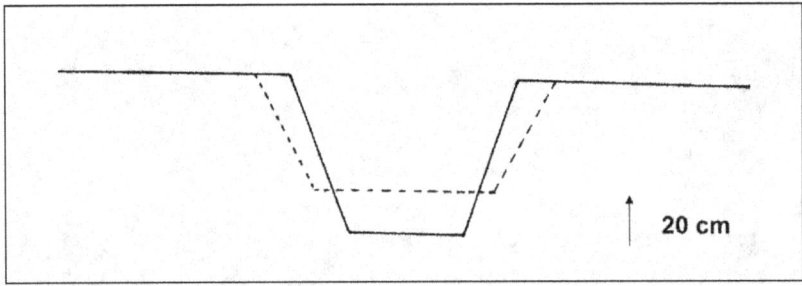

Figure 46: Principle sketch of a channel cross section: During a restoration project in Holstebro Municipality the channel bottom was rose 20 cm, while the discharge cross section was widened at the same time.

3.6.3 Ochre abatement projects within clearly demarked valleys

Rodå was one of the first projects done in advance of the Ochre Action Plan of 1994. It had been supposed to be a valley project with change to extensive agricultural land use. The stream is targeted as B3 upstream and B1 downstream the project area. The valley itself is an ochre potential area (red = class I) between the villages Rødekro and Hellevad. It is well demarked from the surrounding landscape with an average depth of 1 – 2 m under terrain. Adjacent areas are mostly sandy (former heath land). The project area consisted of 5 km stream length and 50 ha valley bottom. Such a project demands detailed surveys of discharge within the whole stream, land survey of the valley and alterations of groundwater table.

Different measures were included in the project. A permanent **pond** (figure 47) with a surface area of 1.2 ha is extended to 4 ha and ½ m depth when the project area is dammed up during winter season. The pond prevents the stream to dry out during summer time. **Damming up** the stream causes a higher ground water table within the aquifers and prevents pyrite and iron to be oxidized and washed out. Additionally stones (with a diameter of 10 – 40 cm) were put on the bottom to **higher the water table** in the channel and adjacent aquifers. The lower boundary of the project area is consistent of an **earth dam**. Free passage is possible because of a by-pass.

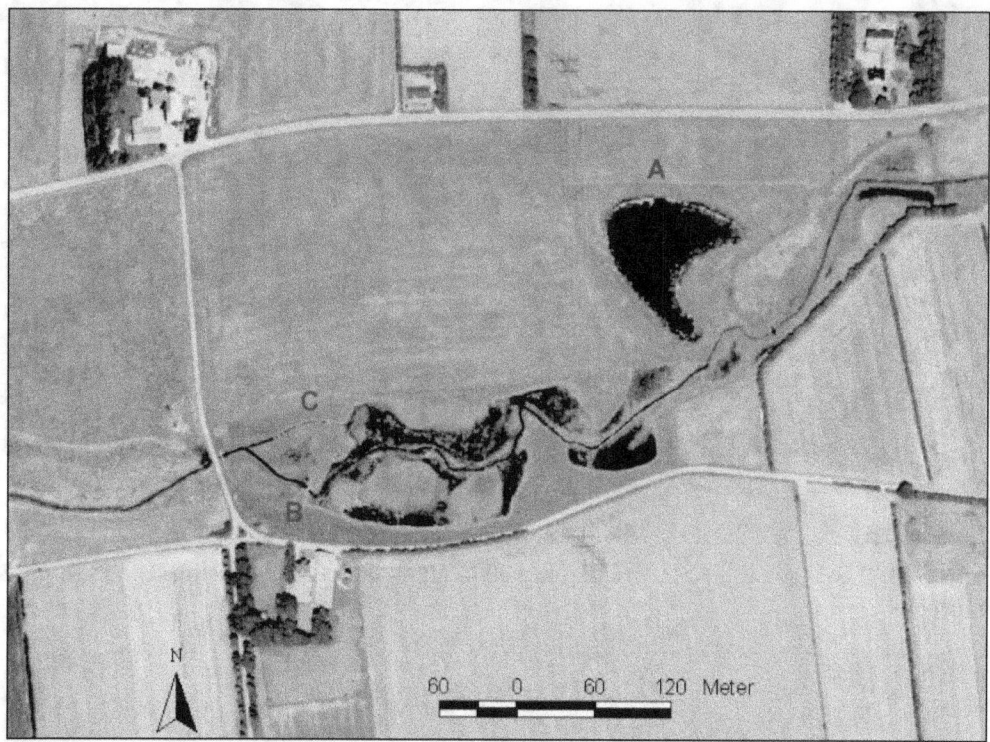

Figure 47: The ortho-photograph of the Rodå project area shows the ochre lake (A), weir (B) and gravel by-pass (C). The stream flows from right to left. [Copyright: Kort & Matrikelstyrelsen (Ministry of the Environment), with permission of the River Section, SJA]

The stream should had been dammed up until 15th April and maintained by mowing weeds manually. Altered maintenance would have caused a higher structural diversity and self cleaning efficiency. The lake has enhanced landscape amenity and provides resting place for birds. It stores also water during high discharge events and prevents flooded areas downstream. Ochre and nutrients are settled in the lake. Public relation and arrangements between work groups were a main part of the project. Compensation was paid for 7 ha, of which 1.5 ha were intensively drained. The duration of the project was restricted to 2 ½ years. With the end of the project stones had to pushed down to lower the water table again (SJA TEKNISK FORVALTNING 1991).

Figure 48 shows the project area at the beginning of May 2005. The flooded valley is not the result of the project anymore but of decayed organic material and a consequently lowered terrain (Moorsackung) within the valley. Actually landowners request to alter the regulative and deepen the channel. Grassland upstream has been changed into arable land, what causes higher discharges (SJA WASTEWATER SECTION, CLAUSEN personal notification 2005). Weed cut does not take place within the project area if sufficient discharge is guaranteed.

Figure 48: The Rodå project area in spring 2004 (flow direction downstream). (Picture: H. Prange, 04/2005)

Figure 49: The Rind Stream (Ringkøbing County). Also this valley could have been dammed up like Løgumkloster Bæk Ochre Lake. (Picture: H. Prange, 05/2005)

The Rind Stream (Ringkjøbing County) is a more successful example of a combination of ochre removal and restoration. It was initiated by the landowners in the beginning of the 1990´s. Shallow basins, covered by grass species that tolerate flooding for a longer time, were excavated within the slopes of the new remeandered stream (figure 49). Ochre is deposited here during winter time. The recommended depth of the basins is also 0.5 m (RINGKJØBING COUNTY, JENSEN personal notification 2005).

3.6.4 Ochre abatement measures within flat terrain

An actual example is the **Nips Stream** in the North of South Jutland County. The Nips Stream is part of the Ribe system, which is one of four catchments areas focused by the National Action Plan for Salmon (KJELLERUP et al. 2004). It aims to improve spawning grounds, removal of obstacles to enable free passage and a better water quality. It is a small tributary to the Gels Stream, which is supposed to be a good spawning ground by the SJA. An "Activity Plan for Self-reproducing Salmon Populations within the Ribe system" will be submitted to the Ministry of Environment (SJA b 2005). This includes a mixture of different measures at the Gels Stream like applications of spawning gravel, restoration and improvement of water quality.

The Nips stream is a small tributary that drains a former wetland. The channel was deepened by intensive maintenance to two metres under terrain. Water quality of the Gels Stream is reduced by ochre from the Nips Stream (dissolved iron concentration is 1–2 mg/l). Initially a valley project was planned in which the water table would have been raised by application of stones and remeandering the channel. But because the terrain is extreme flat an area of **400 ha** would have been **influenced by the altered groundwater conditions**. The stakeholders are not cooperative because the agricultural land is of relative high value. Instead of the valley project it has been suggested to establish five small ochre lakes at the influent to the Gels Stream. Ochre lakes are an alternative solution and just one part of bigger projects that aim good chemical water quality as well as good structural quality and guarantee a sufficient discharge at the same time (SJA a 2005 and SJA RIVER SECTION personal notification 2005).

- Winter-Ockerseen sind überschwemmte Flächen, auf denen Ocker und Nährstoffe sedimentiert werden. Im Sommer werden sie beweidet.
- Seen bekämpfen mehr oder weniger die Symptome, während das Wiederanheben des Grundwasserflurabstandes den Grund für die Verockerung bekämpfen würde.
- Während in deutlich abgegrenzten Autälern weniger Anlieger von Wasserstandsanhebungen betroffen sind, müssen in weitläufigen Gebieten alternative Maßnahmen (wie z.B. Ockerseen) ergriffen werden, da Entschädigungszahlungen die Projekte sehr teuer machen könnten bzw. betroffene Landbesitzer evtl. nicht kooperationsbereit sind.
- Ockerseen sind zumeist Teil größerer Projekte, welche unter anderem eine verbesserte Hydromorphologie und ökologische Durchgängigkeit zum Ziel haben.

3.6.5 Alteration of maintenance

3.6.5.1 Ochre pollution because of hard maintenance with mow baskets

Selective manual weed cut has been proved to cause turbulent channels and stabile edges. It enhances structural diversity and consequently self-cleaning efficiency of streams in a long-term view. **Intensive maintenance** with mow baskets is said to intensify diffuse ochre pollution in different ways:

- Machines can not work accurately. More stones and sediments are removed. Less structural diversity enhances the velocity.

- High velocities within straightened streams make the bottom deeper (especially within sandy areas). The ground water table within adjacent areas is lowered and pyrite exposed to aerobic conditions.
- Complete removal of vegetation in one step causes suddenly lowered water tables and consequently more oxidized pyrite.
- Ochre that has been already trapped is released most extremely when using mow baskets.
- The mow basket disturbs non-oxidized earth layers in the channel.

The last two points have been investigated by Holstebro Kommune during last October at the Savstrup Stream. **Influence of maintenance by mow basket**, by boat with mow bar and with scythe (manual weed cut) on the water chemistry has been investigated. Oxygen, pH, total iron and dissolved iron were measured all five minutes downstream the maintained stretch. Results have been provided as shown in figure 50 and 51. Maintenance with scythe did not release a lot of total iron (ochre) in comparison to the mow basket (figure 50). The peak of total iron caused by the boat with mow bar can be explained as following. The boat was working upstream and following downstream. Amounts of released iron were summed up. Figure 51 shows that maintenance with the mow basket increased concentrations of ferrous iron from 0.7 mg/l to 1.4 mg/l. Concentrations of toxic ferrous iron were doubled by this maintenance measure.

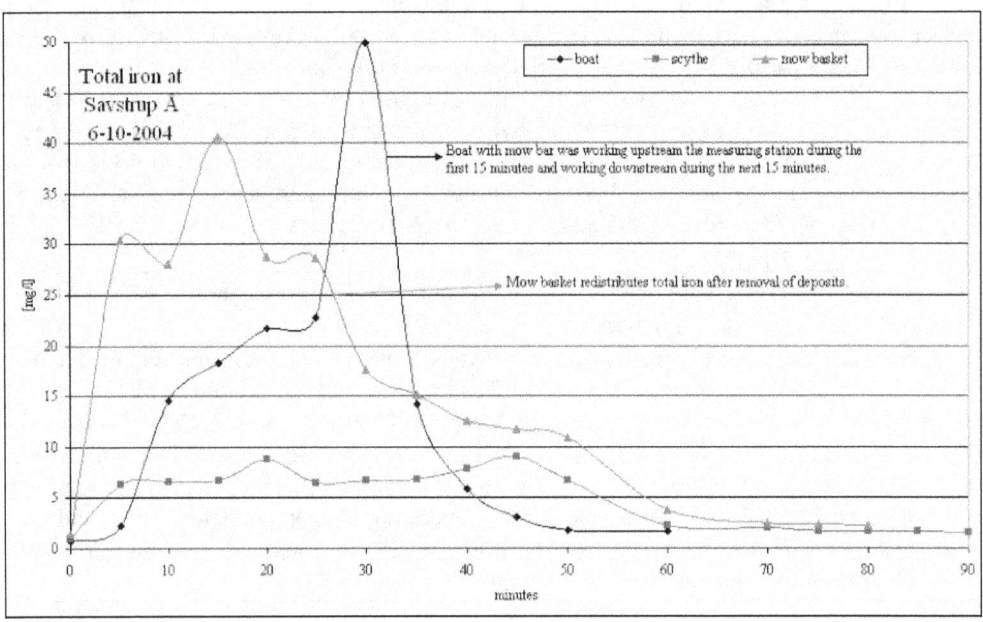

Figure 50: Influence of three different maintenance measures on the concentration of total iron within Savstrup Stream in autumn 2004. (Results have been provided by Holstebro Municipality)

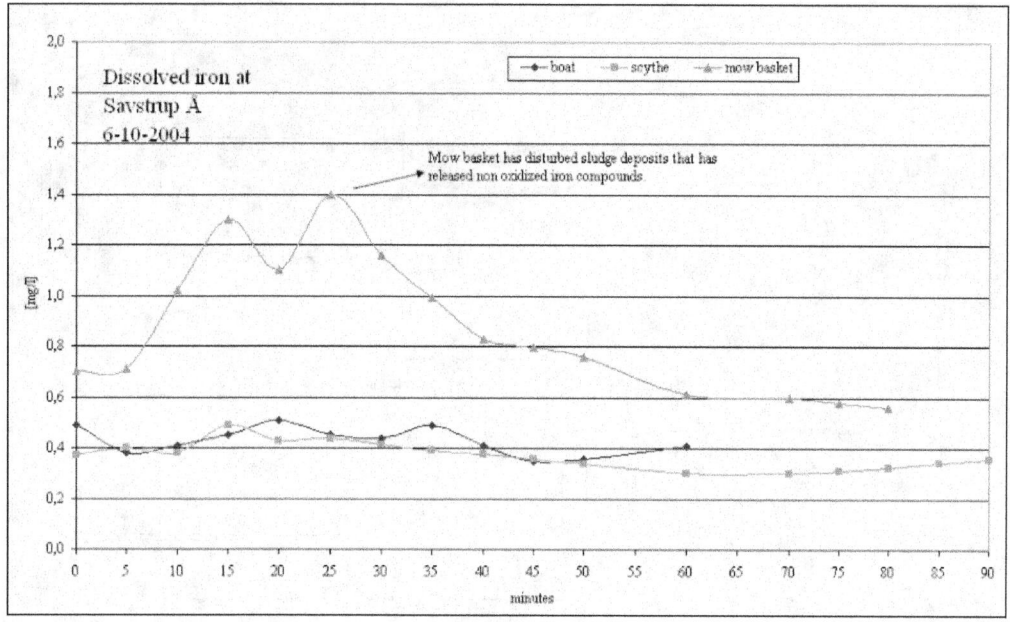

Figure 51: Influence of three different maintenance measures on dissolved iron concentrations at Savstrup Stream in autumn 2004. (Results have been provided by Holstebro Municipality)

3.6.5.2 Manual weed cut

Different counties and municipalities have been interviewed during 1998 to collect information about **maintenance of ochre polluted streams** (CHRISTENSEN & MARCUS 1998). Following paragraphs will summarize the findings and add some information. One conclusion was that weed cut should not be done, if sufficient drainage is guaranteed. Vegetation increases pH and oxygen concentrations. This accelerates oxidation of ferrous iron and sedimentation of ochre. Well developed vegetation causes low ferrous iron concentrations even until spring time. If maintenance is demanded following advices should be considered.

Maintenance of private or small streams should preferably take place during late summer (August), when the discharge is low. Best would be to dam the stream and prevent ochre to be transported downstream. **Weed cut** should take place **more than once** and if possible preventive. This means earlier than it would be necessary to protect adjacent (agricultural) areas against flooding. It is recommended to cut weeds two times per summer season. One early cut in June/July. Another cut should take place before 1st September. This enables the vegetation to establish properly before the winter season. Removal of huge vegetation amounts would lower the water table and expose pyrite to aerobic conditions. Cutting weeds in small steps will cause a higher and **constant water table**. Vegetation should ideally be removed with a scythe and local ochre deposits with a shuffle. This does not disturb the ground or distribute the ochre extremely.

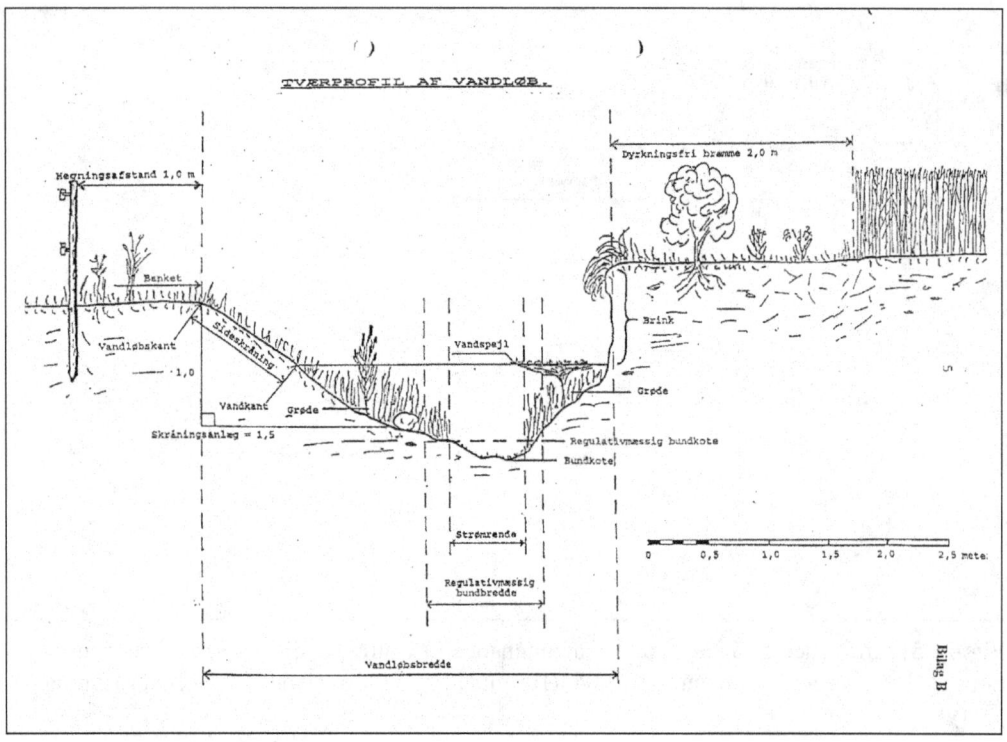

Figure 52: This sketch is usually included in the regulatives. If channel cross sections are deeper and wider as described in the regulatives, weeds can be left in the channel to raise the water table again.

Cross sections are sometimes deeper than described in the regulative. This offers the possibility to leave vegetation at the edges and isles within the channel. The sketch (figure 52) can be found in every regulative. Ochre is deposited within the weeds and becomes more stabile sediment (Verlandung). Vegetation is also a valuable habitat for fish and invertebrates. The channel becomes **thinner, deeper and more turbulent**. The edges are less stressed by erosion. It has been also annotated that after a few years channels are to thin to leaf further vegetation at the edges. The additional application of stones (Querriegel) would prevent channels to be deepened extremely (TENT written notification 2005).

Figure 53 shows weed cut at Fugl Sang Sø in June 2005, Herning Municipality. Weed cut should follow the already existing channel that was formed by the water current. The channel should not be wider than 2/3 of the channel bottom. *Potamogeton natans* needed to be removed here already in early June to guarantee sufficient drainage. The weeds are collected downstream at a rake. Weed cut takes place two times per year (June and October). Within some municipalities vegetation is removed up to four times per year (HERNING MUNICIPIALTY, BRANDT personal notification 2005).

Figure 53: Weed cut at Fugl Sang Sø, Herning Municipality. (Picture: H. Prange, 06/2005)

Selective weed cut means to favour special species that are **winter green** and resistant to erosion on the one hand. Dominating species needed to be reduced on the other hand. For example *Sparganium* spec. can occur in dense stands. To favour the abundance of wintergreen weeds vegetative parts of the plants can be fixed with a stone. Plants will distribute themselves downstream. Plant species in table 14 have been selected because they are winter green. Ideally the complete plant is left during winter time and offers big surface area. This would benefit the oxidation and deposition of iron and ochre what occurs most extremely during winter time. Also remnants of other species like *Phragmites australis*, *Glyceria maxima*, *Phalaris arundinacea* favour sedimentation along the channel edges.

Table 14: Winter green species that enhance cleaning efficiency (based on TENT 2001)

Species Latin Name	Influence
Ranunculus spec.	Should not be cut if possible, valuable resting place for fish, strong stands can possibly occur during **winter time**, big surface area
Berula erecta	Weed cut is not necessary, submerged leafs during **winter season**
Elodea Canadensis	Dense stands, strong current within freshly cut channels prevents *Elodea* spec. to overgrow it again and the embankments are stabilized, **winter green** but only small stands are left
Callitriche spec.	Should not be cut if possible, is well developed during **winter season**, big surface area
Nasturtium microphyllum	Weed cut in autumn is usually not necessary, submerged leaves are **winter green**

- Die intensive Gewässerunterhaltung führt zu Grundwasserabsenkung und verschlimmert damit die diffuse Verockerung.
- Während Mähkorbeinsätzen werden unoxidierte Erdschichten angeschnitten, was zu einer Erhöhung der Konzentration von gelöstem Eisen und zur Sauerstoffzehrung führt.
- Durch die maschinelle Unterhaltung wird bereits sedimentierter Ocker wieder freigesetzt.
- Die gezielte Pflanzenmahd von Hand würde die Struktur der Gewässer verbessern und damit ihre Selbstreinigungskraft erhöhen.
- Komplette Entfernung der Vegetation lässt den Wasserstand in angrenzenden Flächen fallen und Pyrit wird oxidiert.
- Eine mindestens zweimalige und schonende Mahd führt zu einem konstanteren Wasserstand.
- Wintergrüne Pflanzen mit großer Oberfläche oxidieren Eisen und sedimentieren Ocker.

4. Discussion

4.1 Know how and experience are already available

The Ochre Act was passed in 1985. This led to different research projects which aimed to determine the negative impact of ochre pollution on the aquatic environment and to identify ochre potential areas, threshold limits for pyrite concentrations within soil as well as threshold limits for ferrous iron within watercourses. Extensive investigations and strategies seemed to be organized in a pragmatic way, e.g. the use of historical maps for the identification of ochre potential areas. Combating ochre pollution is only one part of the work within counties and municipalities in Jutland, but its abundance and possible abatement measures have been **involved in many projects** during the last decade.

A long-term **monitoring** of ferrous and ferric iron concentrations within watercourses is necessary to identify location of sources as well as to observe development of ochre pollution. Although the mapping of ochre potential areas was revised in the early 1990´s strategies as described under 3.1 and following chapters can provide valuable experiences to tackle ochre pollution within N/W-Germany where it has not been recognized as an **ecological problem of the aquatic environment**, yet. Administration and approval of drainage projects is not necessary anymore but nevertheless reasons of ochre pollution within the catchments area need to be identified to combat ochre pollution in the most effective way.

> - Das Ocker-Gesetz ist 1985 erlassen worden und Maßnahmen zur Bekämpfung der Verockerung sind seit ca. zehn Jahren fester Bestandteil verschiedener Projekte.
> - Das Langzeit-Monitoring von Konzentrationen zwei- und dreiwertigen Eisens, als auch die Identifikation von Ursachen im Einzugsgebiet, sind für effektive Gegenmaßnahmen notwendig.
> - Bereits in Jütland gemachte Erfahrungen können auf N/W-Deutschland übertragen werden.

4.2 Investigation of the cleaning efficiency of three ochre lakes in SJA

4.2.1 Impact of the catchments area on amounts of iron that reach the ochre lake

While investigating the cleaning efficiency of ochre lakes it needs to be considered how much ferrous iron or ochre reaches the lake. This factor is mainly determined by the characteristics of the catchments area (geology, land use, hydrology and anthropogenic alterations) which influences physico-chemical parameters and amounts of iron that are leached out of the soil and into the recipients. High oxygen concentrations and pH accelerate the oxidation of ferrous iron to ferric iron. High alkalinity prevents extreme alterations of the pH caused by ochre pollution. Physico-chemical parameters influence therefore **intensity and phenotype** of ochre pollution. Within watercourses of low oxygen content and pH toxic ferrous iron is transported further downstream than in watercourses where ferrous iron is oxidized fast.

Ferrous iron has to be oxidized and settled, while ferric iron or ochre needs to be sedimented only. Iron concentrations that reached the three ochre lakes were nearly the same. The average concentration of dissolved iron was a bit higher at Landeby Bæk Ochre Lake than at the other two lakes (2.0 mg/l in comparison to 1.8 mg/l). Concentrations of dissolved iron are positively correlated with the discharges as well as with daily precipitations at Landeby Bæk Ochre Lake (Figure 37). Increased concentrations of ferrous iron because of higher discharges were also proved at the other two ochre lakes (figure 12 and 26). The impact of even daily precipitation on ferrous iron concentrations might be explained by the **small catchments area** of Landeby Bæk Ochre Lake, which is only 3 km². Also surprising was the **positive relation between pH and dissolved iron** (figure 43). Usually high pH accelerates oxidation of ferrous iron (negative relation). Maybe lime fertilizers as well as ferrous iron are leached from the adjacent areas during strong precipitation events. Landeby Bæk Ochre Lake is located in an ochre potential area and ferrous iron has not been oxidized when it reaches the lakes inlet.

The Hvirlå Ochre Lake was installed downstream two former wetlands and would also receive ferrous iron. Ferrous iron concentrations are negatively correlated with the oxygen concentrations at the lakes inlet. The **oxygen** concentrations are usually **high in the upper Hvirlå stream** (ANDERSEN et al. 1995). This accelerates oxidation of ferrous iron. No correlation could be found between physico-chemical parameters and dissolved iron at Løgumkloster Bæk Ochre Lake. In comparison to the Hvirlå Ochre Lake Løgumkloster Bæk receives higher ochre loads from several small tributaries. The position of the ochre lake in the stream system is therefore an important factor that influences the cleaning efficiency. The **negative relation between dissolved iron and temperature** proves that ochre pollution occurs more intensive during winter season (figure 22, 32 and 42).

> - Das Einzugsgebiet der Seen bestimmt physikalisch-chemische Parameter, die Mengen an ausgewaschenem Eisen und damit auch die Intensität der Verockerung.
> - Das Einzugsgebiet und die Lage des Ockersees im Gewässersystem beeinflussen die Eisen- bzw. Ockerfrachten am Einlauf des Sees und damit auch die Reinigungseffizienz.
> - Das Einzugsgebiet des Landeby Bæk Ockersees ist verhältnismäßig klein. Konzentrationen gelösten Eisens werden sogar von täglichen Niederschlägen erhöht.
> - Der Hvirlå Ockersee wird auf Grund seiner Position unterhalb eines ehemaligen Feuchtgebietes auch von gelöstem Eisen erreicht, jedoch ist die Sauerstoffversorgung des Gewässers gut, was die Oxidation des Eisens vorantreibt.
> - Der Løgumkloster Bæk Ockersee sammelt die Frachten verschiedener kleiner Zuläufe. Beziehungen zwischen Konzentrationen gelösten Eisens und physikalisch-chemischen Parametern konnten hier nicht festgestellt werden.

4.2.2 Reasons for the different cleaning efficiencies within the investigated ochre lakes

4.2.2.1 Influences on the improved cleaning efficiency at the Hvirlå Ochre Lake

Cleaning efficiencies of the three investigated ochre lakes are different. It is influenced by several factors, of which the retention time is one of the most important ones. The enlargement of the Hvirlå Ochre Lake took place in the year 2000 and aimed longer retention times because of a bigger volume. Since the same year concentrations of **dissolved iron decreased upstream** the lake. The improved cleaning efficiency is therefore not only caused by a bigger volume and consequently longer retention times but also by less dissolved iron that has reached the lake.

Theoretically the enlargement would have doubled the retention time. The middle part of the lake is nearly completely overgrown by *Potamogeton natans* (figure 18). Excavation of earth and sediments from the northern side of the lake during the enlargement might have favoured the development of the **channel.** It reduces the total lake width of 104 metres to an effective width of only 10 metres and consequently reduced retention time and catalytic surface. The channel diminishes the improvements caused by the enlargement. Instalment of panels within the shallow part of the ochre lake could possibly cause a more regular distribution of the current on the total width of the lake and lead to a regular distribution of vegetation and consequently improved cleaning efficiencies.

The ochre layer within the lake was initially assumed to be thicker especially in comparison to the huge amounts of **258 t** deposited ochre that have been calculated for the last 12 years (Table A4d). A possible explanation could be that the lake was excavated deeper than described in the project proposals.

- Die verbesserte Reinigungsleistung des Hvirlå Ockersees lässt sich durch die im Jahr 2000 erfolgte Erweiterung, aber auch durch stromaufwärts auftretende geringere Eisenkonzentrationen erklären.
- Die Stromrinne, verursacht durch dichte Bestände des *Potamogeton natans* im Abschnitt 2 des Ockersees, beeinflusst die Aufenthaltszeit erheblich.

4.2.2.2 Influences on the cleaning efficiency of Løgumkloster Bæk Ochre Lake

Especially cleaning efficiency of total iron seemed to decrease in 1999 and 2004 in this lake. It is hard to say if this is a permanent trend because only a few data were available here. The **old channel** has favoured the distribution of vegetation in the upper part of the lake. Several channels in the compartments one to three (see figure 28) reduce the retention time of water within this ochre lake. Islands occur especially within compartment 4 (figure 54). Another problem is the **missing sand trap** at the inlet. Deposits consist of ochre but also of sand and organic material. Especially if the water table reaches only heights of 10.45 m DNN at low discharges, the calculated volume of deposits has reduced the volume extremely from approximately 13.700 m³ to only 2.000 m³ (table 11).

Figure 54: Several islands occur within compartment 4 at Løgumkloster Bæk Ochre Lake (compartments are shown in figure 28). (Picture: H. Prange, 06/2005)

Planned improvements involve the elevation of the dam to 11.00 m DNN, the excavation of the old channel within the upper part of the ochre lake (compartment 1 to 3, figure 28) and the instalment of a sand trap at the inlet of the ochre lake. Theoretically the elevation of the water table would prolong the retention time to approximately 20 hours (assuming 950 l/s, what is the 95 % percentile). This would be more as **double as long as the recommended** eight hours. But it has to be considered that adjacent shallow areas are already overgrown by vegetation.

Abundance of vegetation within ochre lakes has always advantages and disadvantages according to the cleaning efficiency of ochre lakes. It can accelerate oxidation and sedimentation of iron but it causes also a higher input of organic material. This leads to reductive conditions and maybe resuspension of already settled iron. **Decay of organic material** reduces also oxygen concentrations what is disadvantageous to iron oxidation as well as the ecosystem. Occasional low oxygen concentrations (< 4 mg/l) occurred especially during summer season. The elevation would cause a **bigger surface-volume-ratio**. Enhanced surface areas would cause more exchange of gases what is benefiting the iron oxidation but also the photo reduction. Wind disturbance and resuspension is unlikely because the valley and surrounding woods offer wind shelter. Especially at this lake it becomes clear that cleaning efficiency is influenced by several factors. Vegetation will continue to overgrow the lake (succession) without any maintenance.

- Die Reinigungseffizienz für Total Eisen ist das Problem dieses Sees. Die sich vom Altlauf ausbreitende Vegetation und der fehlende Sandfang führten zu einer starken Verringerung des Volumens und damit zu verkürzten Aufenthaltszeiten.
- Der hohe Anteil an organischem Material verursacht gelegentlich Sauerstoffzehrungen, was die Eisenoxidation, sowie das Ökosystem negativ beeinflusst.
- Die geplante Erhöhung des Damms auf 11,00 m DNN würde die Retentionszeit um mehr als das Doppelte der empfohlenen Retentionszeit von acht Stunden erhöhen.
- Das Oberflächen-Volumen-Verhältnis wird sich durch die Erweiterung erhöhen. Dies ist positiv hinsichtlich der Oxidation von gelöstem Eisen, kann aber auch eine verstärkte Photoreduktion verursachen.

4.2.2.3 Reasons for the good cleaning efficiency of Landeby Bæk Ochre Lake

The main reason for the extreme good cleaning efficiencies in this lake of more than 90 % for dissolved iron is a **low discharge** and consequently **long retention times** of approximately 40 hours (Table A36). Cleaning efficiency of total iron decreases regularly during winter seasons (figure 35). The Landeby Bæk Ochre Lake is exposed to wind. Surrounding plantings at the Western side of the ochre lake have not developed sufficiently, yet. Strong influence of wind might accelerate the oxidation of dissolved iron but also cause resuspension of ochre particles. Occasional high precipitation and discharge events wash ochre out of the lake. This explains why the amount of dissolved iron deposits is bigger than total iron deposits (Table A31c).

The vegetation is also responsible for the strong reduction of dissolved iron. Figure 55 and 56 show the difference between water at the inlet and outlet of the ochre lake. A strong abundance of *Callitriche* spec. has already been noted during a site visit in August 2001 (BALLEBY 2001). ***Callitriche* spec. is a winter green** plant and favours oxidation of dissolved iron. During winter season only the stems of *Potamogeton natans* are left in the deep part of the lake. It takes up carbondioxide (CO_2) at its whole surface. An increased pH and higher oxygen contents favour oxidation of iron. Dense stands of *Potamogeton natans* offer also space for sedimentation.

***Chara* spec**. was noted especially within the shallow areas at the edges of the lake during the investigations on 22nd June. It exchanges also gases at its whole surface and its branches offer a huge surface area for oxidation and sedimentation. A strong abundance of *Chara* spec. in June and July was also observed within Herning Municipality (THORDAL-CHRISTENSEN 2004). It supports cleaning of iron and ochre during summer season and is also an indicator for good water quality. The shape of the lake seems not to influence the cleaning efficiency but maybe causes a regular distribution of the current. Although sedimentation is more intensive at the edges of the lake depositions seems to be distributed across the whole width (figure 39). Landeby Bæk Ochre Lake is one of the newer lakes but deposits have already reduced the retention time by 40 % here (table A36).

Figure 55: *Callitriche* spec. covered by ochre at the inlet of Landeby Bæk Ochre Lake.(Picture: H. Prange, 06/2005)

Figure 56: This picture was taken close to the outlet of Landeby Bæk Ochre Lake. The water was absolutely clear and *Chara* spec. occurred in dense stands. (Picture: H. Prange, 06/2005)

- Hauptursache für die extrem gute Reinigungseffizienz des Landeby Bæk Ockersees (ca. 90 % für gelöstes Eisen) sind die langen Retentionszeiten.
- Auch das starke Auftreten von *Callitriche* spec. (wintergrün) und *Chara* spec. treibt die Eisenoxidation voran.
- Beeindruckend war der Unterschied zwischen dem getrübten Wasser im Einlauf und glasklarem Wasser am Auslauf des Ockersees.
- Eine besonders in der Wintersaison auftretende Verringerung der Reinigungsleistung von Total Eisen wird durch starke Abflussereignisse verursacht.

4.2.2.4 Possibilities to improve the cleaning efficiency of ochre lakes

As described within chapter 3.2.6 decreasing cleaning efficiencies of ochre lakes during winter season and possible improvements are the main purpose of recent research projects. The application of artificial substrates and Christmas trees was too expensive and not efficient enough. It is therefore recommended here to **favour winter green plants** during the installation of new ochre lakes. Instead of using common grass seeds it is maybe possible to **incorporate parts of aquatic or swamp plants** in the top soil, apply it on the lakes bottom and flood it directly. Most aquatic plants reproduce themselves vegetative and maybe weeds from manual weed cut can be used. If possible, only weeds from the same watercourse should be used to prevent the transfer of fish diseases. Different plant species try to enlarge their surface area within standing waters (Heterophylie) for a more efficient uptake of carbondioxide and better photosynthesis efficiency (SCHÖNBORN 2003). This would also cause a bigger active surface and benefit iron oxidation as well as sedimentation.

- Besonders die Verbesserung der Reinigungsleistung der Ockerseen in der Wintersaison ist Ziel aktueller Forschungsversuche.
- Künstliche Substrate zur Erweiterung der aktiven Oberfläche hatten sich als zu zeitaufwendig und kostspielig erwiesen.
- Das Einbringen wintergrüner Makrophyten (aus dem gleichen Gewässersystem) in die Ockerseen sollte weiter erforscht werden.

4.2.3 Consequence for the aquatic ecosystem downstream the ochre lakes

4.2.3.1 Altered physico-chemical parameters downstream the ochre lakes

Oxygen concentrations and pH decreased over years upstream Hvirlå Ochre Lake and Løgumkloster Bæk Ochre Lake, while loads of total iron increased at the same time (chapter 3.5.1.2 and 3.5.2.2). Without the instalment of the ochre lakes ochre pollution would have continued or become worse. While the watercourse might provide acceptable living conditions for fish during summer season it is the question what would live or spawn here if the stream would be still polluted by 200 – 300 kg/day total iron during winter season?

Hvirlå and Landeby Bæk Ochre Lake **increased oxygen and pH** in average, while Løgumkloster Bæk Ochre Lake decreased it. **Higher temperatures** (5 – 7 °C) occurred rarely during summer season at Løgumkloster Bæk and Landeby Bæk Ochre Lake (Table A26 and A39). In general alteration of temperature itself is not a problem for organisms but it can influence other parameters like oxygen supply. The Hvirlå has been targeted as cyprinoids stream until the village Ravsted. Optimal **oxygen** supply for **cyprinoids is > 5 mg/l** (SCHRECKENBACH 2005). Oxygen supply has been sufficient in the stream and has been even increased by the ochre lake. Løgumkloster Bæk is a **salmonoids stream**. The optimum oxygen concentration is **7 mg/l** (SCHRECKENBACH 2005). The oxygen concentration within Løgumkloster Bæk seemed to decrease in general what can be intensified by the decay processes within the ochre lake during summer season. Oxygen concentrations have been increased by more than 2.5 mg/l at Landeby Bæk Ochre Lake which is a small tributary without stream target but enters a salmonoids stream. The natural high alkalinity prevented extreme alterations of the pH in the lakes.

> - Sauerstoffkonzentrationen und pH stromaufwärts des Hvirlå und des Løgumkloster Bæk Ockersees sanken über Jahre hinweg. Ohne das Einschalten der Ockerseen hätte sich dieser Prozess vermutlich auch weiter stromabwärts fortgesetzt.
> - Erhöhte Temperaturen (5 – 7°C) kamen vereinzelt in der Sommersaison im Løgumkloster Bæk und Landeby Bæk Ockersee vor.
> - In den meisten Fällen erhöht die Vegetation in den Ockerseen die Sauerstoffkonzentrationen auf ein für die Gewässerzielsetzungen positives Maß.
> - Die natürlich hohe Alkalinität verhindert extreme Veränderungen des pH-Wertes.

4.2.3.2 Improved living conditions for flora and fauna

The average winter concentrations of **dissolved iron** at the outlets of two of the three ochre lakes are still **higher than the demanded threshold limits**. But it should be considered that ochre lakes are often a part of a bigger project. The aim is to reduce ochre pollution in the whole stream system (e.g. the river Brede) and every ochre lake makes its contribution. The WFD emphasizes the ecosystem function of lakes and rivers as habitat (UBA 2004). Good status means that species composition and abundance show only slight alterations as a result of anthropogenic impacts on physico-chemical and hydromorphological parameters (Appendix V, Number 1.2.1 of the Directive 2000/60/EG). Ochre pollution has a negative impact on both. As shown during the investigations pH and oxygen concentrations are decreased and interstitial gap systems are clogged. Ochre abatement measures, which are mostly ochre lakes, led to an **improvement of biological water quality at 48 km** of initially 153 km stream length in the SJA, where a determination of the biological water quality was not possible because of the extreme ochre pollution (SJA a 2002). Improvements and reinstallation of populations need time and monitoring of benthic invertebrates would be best to evaluate the improvements in a long term view.

Ochre lakes are often said to prevent **free passage**. Especially during winter season streams are that extremely polluted that no passage would occur. The parts upstream the ochre lakes are often lost as living space. At the three investigated ochre lakes free passage for fish seems not to be a problem. Initially installed panels at in- and outlet for the distribution of the current have already been rotted. Passage of benthic invertebrates seems to be unlikely because of the thick ochre layers.

> - Die Verockerung verschlechtert sowohl physikalisch-chemische als auch hydromorphologische Qualitätsparameter.
> - Die Ockerseen haben die biologische Gewässergütebestimmung in bestimmten, zuvor stark verockerten, Gewässerabschnitten überhaupt erst möglich gemacht.
> - Verbesserungen brauchen Zeit und gutes Monitoring.
> - Wenn die Ockerseen hinsichtlich der Durchgängigkeit kritisiert werden, sollte geprüft werden, ob überhaupt Wanderung unter verockerten Umständen stattfinden würde.
> - Generell scheinen die dicken Ockerschichten jedoch zumindest die Durchgängigkeit für benthische Wirbellose zu blockieren.

4.3 Combat reasons rather than symptoms – an outlook

4.3.1 Ochre lakes in comparison to other measures

Although ochre appears naturally at wells and springs it has been proved as an ecological problem of the aquatic environment. Structural as well as physico-chemical parameters are influenced negatively and habitats are destroyed where ochre is leached from the soil into watercourses. While some decades ago ochre pollution was mostly caused by drainage projects the Danish Ministry of the Environment came to the conclusion that ochre pollution occurs mostly **diffuse** and is the consequence of intensive or even illegal maintenance as well as by drainage projects that took place years ago. Recent ochre pollution is mostly **permanent**.

Ochre lakes are one possible measure to diminish ochre pollution. Its instalment is the first choice if ochre pollution is caused by small, non-targeted drainage ditches (point sources). Ideally several small ochre lakes should be positioned as high as possible in the stream system rather than one big ochre lake downstream. This would lengthen improved watercourses and enable fish and benthic invertebrates to access streams further upstream. Although ochre lakes seem to be the cheaper and easier possibility, it needs to be considered that **ochre and other sediments** from the sand traps **have to be disposed**, what can also be expensive.

While ochre lakes combat symptoms **winter ochre lakes** can cause higher groundwater tables during winter season and prevent iron to be washed out into surface waters. During summer season it can be still possible that groundwater tables are lower and pyrite is exposed to aerobic conditions. As supporting measure manual weed cut for at least two times per summer season could cause a more constant water table. It should be also considered if deposited ochre compounds on the meadows can be taken up by plants and life stock and finally enter the food chain. On the other hand projects like the described Ravsted Winter Ochre Lake (chapter 3.6.1) favour special nature types.

If the terrain is very flat raising the groundwater tables would have an impact on huge areas as it was the case at the Nips Stream (described under 3.6.4). Ochre lakes are the possible alternative. Also the examples of the Rind Stream and the Rodå Valley Project (described within chapter 3.6.3 and 3.6.4) made clear that the choice of applied ochre abatement measures is influenced by acceptance and economic interests of the stakeholders and compromises need to be found. Recreation and hunting are important arguments. Nevertheless **restoration of wetlands** and raising water tables again would be the best solution and keep the iron in the soil. Water bodies and their catchments areas need to be accepted as one unit in which groundwater, surface waters and flood plains interdependent to each other (UBA 2004).

Another **important point to combat (especially diffuse) ochre pollution** is to reduce intensive maintenance and try to favour winter green plant species by **selective manual weed cut** (chapter 3.6.5). Danish experiences show that sufficient drainage can be guaranteed while enhancing structural diversity and self cleaning efficiency. An altered maintenance would diminish the release of ferrous iron as well as ochre in the watercourses and enhance landscape amenity at the same time.

- Verockerung ist ein gewässerökologisches Problem das Lebensräume zerstört.
- Die heute auftretende Verockerung ist zumeist diffus und permanent.
- Ockerseen sind die anscheinend kostengünstigere und einfachere Lösung, jedoch muss auch deren langfristige Unterhaltung und Entsorgung des Ockers bedacht werden.
- Winter-Ockerseen verhindern die Auswaschung des Eisens während der Wintersaison, jedoch sollte überprüft werden, in wie fern Ocker und andere abgelagerte Stoffe durch die Beweidung im Sommer in die Nahrungskette gelangen können.

> - Weitgreifende Maßnahmen können an mangelnder Akzeptanz und wirtschaftlichen Interessen scheitern.
> - Trotzdem sollte es primäres Ziel sein die Ursachen der Verockerung durch Wiedervernässung und eine nachhaltige Unterhaltung zu bekämpfen.

4.3.2 Application of already abundant know how?

Raising public awareness and try to enhance understanding of processes are necessary to receive acceptance for ochre abatement measures. Landscape development and countryside change in **N/W-Germany** are similar to that in Jutland and pyrite is abundant everywhere. **Ochre pollution** has also been described as **increasing in several watercourses** e.g. close to Delmenhorst (Lower Saxony) because of channel regulation and following lowered groundwater tables (LRP DELMENHORST 1998). Figure 57 shows the Immer Bäke close to the city Delmenhorst in comparison to a ditch in the Vide catchment (figure 58).

Lower Saxony (and Mecklenburg-West Pomerania) has provisionally designated water bodies which have been morphologically altered for the main use of water abstraction in farmland as "Heavily Modified Water Bodies" (**HMWB**) and will have to fulfil only the "good ecological potential" (UBA 2004). Intensity of ochre pollution in N/W-Germany needs to be investigated and monitored (as it is demanded by the WFD for the implementation of the monitoring programmes until 2009) to identify ochre sources and combat or at least diminish it. One main aim of this dissertation work was to **summarize and translate** already abundant know how and experiences from Jutland, Denmark. The provision of information as well as excursions to several ochre abatement projects enabled the description of different measures, which can be applied and maybe **improved in international cooperation** now.

> - Auch in Niedersachsen wird die Verockerung als ein sich verstärkendes Problem auf Grund der Gewässerregulierung und intensiver Unterhaltung beschrieben.
> - Viele der zur Entwässerung bestimmten Gewässer sind vorläufig als „erheblich veränderte Wasserkörper" eingestuft worden, dessen Ziel das „gute ökologische Potential" ist.
> - Die Verockerung sollte im Rahmen der Durchführung der Wasserrahmenrichtlinie in Monitoring and Maßnahmenprogrammen einbezogen werden.
> - Das zur Verfügung gestellte Material und Exkursionen haben die Beschreibung verschiedener Maßnahmen zur Bekämpfung der Verockerung in Jütland möglich gemacht. Diese können nun auch in N/W-Deutschland angewandt und evtl. in internationaler Zusammenarbeit verbessert werden.

Figure 57: The Immer Bäke close to the city Delmenhorst, Lower Saxony. (Picture: H. Prange, 10/2004)

Figure 58: Ditch within the Vide catchments area, May 2005. (Picture: H. Prange, 04/2005)

5. Literature

ANDERSEN, M.B., BACH, H., MADSEN, M., MØLLER, B. (1995): Okkermodel for Hvirlå – En Analyse af Belastning og Indgrebsmuligheder [Ochre model for Hvirlå – An analysis of loads and possible combat measures]. Rapport til Sønderjyllands Amt, Sagsnr. 755336.

BALLEBY, C. (2001): Investigation of deposition and vegetation cover within the ochre lakes of the Sønderjyllandsamt, Tønder.

BAUR, W. H. (1997): Gewässergüte bestimmen und beurteilen. 3., neu bearbeitete Auflage, Parey Verlag, Berlin, 209 S.

BREHM, J. & MEIJERING, M.P.D. (1996): Fließgewässerkunde – Einführung in die Ökologie der Quellen, Bäche und Flüsse. 3. überarbeitete Aufl., Quelle und Meyer (biologische Arbeitsbücher), Wiesbaden, 302 S.

CHRISTENSEN, L.B. (1992): Miljøprojekt Nr. 192 – Dimensioniering af grødefyldte bassiner til okkerrensning [Dimensioning of weed filled basins for ochre purification]. - Det Danske Hedeselskab, Forsogsvirksomheden, Viborg.

CHRISTENSEN, L.B., MARCUS, E. (1998): Okkerundersøgelser – Forskrifter for vendligholdelse af okkerbelastede vandløb [Ochre investigations – Instructions for maintenance of ochre polluted streams]. Sag nr. 132–97065, Hedeselskabet Viborg.

CHRISTENSEN, L.B., OLESEN, S.E. (1985): Hedeselskabet: Leaching of Ferrous Iron after Drainage of Pyrite-rich Soils and Means of Preventing Pollution of Streams (85-23) – Saertyk af Proceedings of a symposium on Agricultural Water Management. Arnhem, Netherlands, 18 – 21 Juni 1985.

CLAUSEN, J. (1994): Okkerhandlingsplan 1994. Spildevandsafdelingen [Wastewater Section]

EGGELSMANN, R. (1973): Dränanleitung - Landbau, Ingenieurbau, Landschaftsbau. Verlag Wasser und Boden, Axel Lindow & Co, Hamburg, 332 S.

[EU-WFD] EU-Waterframework – Directive 2000/60/EG of the European Parliament and of the Council of 23rd October 2000 establishing a framework for Community action in the field of water policy.

GRIEBLER, C. & MÖSSLACHER, F. (2003): Grundwasser – Ökologie, Fachatlas. UTB-Verlag, Wien, 495 S.

GRØN, P.N. (2000): Restauringen af Bredeå 1991–1998 – Redegørelse for restauringens effect på nature og miljø i aen og det omgivende terrain [Restoration of the river Brede 1991 – 1998 – Evaluation of restoration effects on nature and environment within valleys and adjacent areas]. Bio/consult, Johs Ewalds Vej 42–44, 8230 Abyhoj.

HASLAM, S.M. (1987): River Plants of Western Europe – The macrophytic vegetation of watercourses of the European Economic Community. Cambridge University Press, Cambridge, London, New York, Rochelle, Melbourne, Sydney, 483 S.

JENSEN, A.R., NIELSEN, H.T., EJBYE-ERNST, M. (2003): National Management Plan for the Houting. <http://www2.skovognatur.dk/udegibelser/2004/Forvaltningsplan_for_snaebel_en gelsk.pdf > (03.06.05)

JENSEN, P.S., KRISTENSEN, L.A., OTTOSEN, O., BRANDT, S., AGAARD, P., KOFOED, F. (2004): Okker. Et vandløbsproblem, vi kann gore nøget ved [Ochre – A problem of watercourses we can cope with]. Ringkjøbing Amt, Ribe Amt, Sønderjyllands Amt, Herning Kommune, Holstebro Kommune < www.okker.dk > (02.05.05)

KJELLERUP LARSEN, L., MADSEN B.L., SIMONSEN, P. (2004): National forvaltningsplan for Laks [National Administration Plan for Salmon]. Miljøministeriet & Skov- og Naturstyrelsen [Ministry of the Environment & Danish Forest and Nature Agency], København.

KOFOED, F. (2004): Vudering af effekten af 9 okkerrenseanlæg i vandløb i Holstebro Kommune [Effect evaluation of nine ochre abatement plants at watercourses within Holstebro Municipality]. Holstebro Kommune, Miljøafdelingen og Miljøcenter Vestjylland [Environmental Section Holstebro Municipality and Environmental Centre West Jutland].

KÖLLE, W., WERNER, P., STREBEL, O. & BÖTTCHER, J. (1983): Denitrifikation in einem reduzierenden Grundwasserleiter. Vom Wasser 61. Band, S. 125 – 147.

KOROM, S.F. (1992): Natural Denitrification in the Saturated Zone: A Review, Water Resource Research, **28** (6), Pages 1657 – 1668; County of Civil and Environmental Engineering, Utah State University, Logan.

KUNTZE, H. (1978): Verockerung – Diagnose und Therapie. Schriften des Kuratoriums für Wasser- und Kulturbauwesen, Heft 32, Parey Verlag, Hamburg und Berlin, 133 S.

KÜSTER, H. (1999): Geschichte der Landschaft in Mitteleuropa – Von der Eiszeit bis zur Gegenwart. 20. – 32. Tsd. Auflage der Ges.-Auflage, Beck Verlag, München, 423 S.

LRP DELMENHORST (1998): Landschaftsrahmenplan der Stadt Delmenhorst – Stand: 15. Juli 1998. AG Landschaftsökologie und Umweltplanung, Postfach 1156, 26205 Hatten-Sandkrug, 210 S.

MILJØSTYRELSEN (1984): Okker - Redegørelse om den tre-ærige forsægsordning til nedbringelse af okkergener i vandløb [Ochre – Evaluation of three years investigations about diminishing ochre pollution within ochre polluted streams]. Bilag 12. Miljøstyrelsen [Ministry of the Environment] København.

MILJØSTYRELSEN (1986): Miljøprojekt Nr. 78 – Drænvandskvalitet fra pyritholdige arealer [Quality of drainage waters within pyrite containing areas]. Miljøstyrelsen [Ministry of the Environment] København.

MILJØSTYRELSEN, SKOV- OG NATURSTYRELSEN, GEUS U. GEOGRAF FORLAGET (2004): Det sydlige Jylland – En beskrivelse af omrader national geolokisk interesse [South Jutland – Description of an area with national geological importance]. 1 udgave, 1. oplage 3000 stk, printed in Denmark.

MØLLER, B., THOMSEN, P.G. (1992): Okkermodel for vandløb, Modul til vandløbsmodellen Mike 11 – Rapport til Sønderjyllandsamt, Tønder. of the Sønderjyllandsamt, Tønder.

OKKERGRUPPEN (1999): Notat om status for okkerhandlingsplanen [Report of the status of the Ochre Action Plan 1994], Sønderjyllandsamt, Tønder.

POTT, R. (1999): Nordwestdeutsches Tiefland zwischen Ems und Weser – Kulturlandschaften: Exkursionsführer mit 9 Exkursionen. Ulmer Verlag, Stuttgart (Hohenheim), 256 S.

POTT, R., REMY, D. (2000): Gewässer des Binnenlandes. Ulmer Verlag, Stuttgart (Hohenheim). 255 S.

PRANGE, H. (2005): Die ökologische Bedeutung der Verockerung und ihre Relevanz für die Umsetzung der EG-Wasserrahmenrichtlinie. Projektarbeit im 7. Semester Umweltbiologie der Hochschule Bremen. < http://www.umwelt.schleswig-holstein.de/servlet/is/54631/ > (11.08.2006)

REKER, L. (1984): Lovgivning om okker [Legislation of ochre]. Hedeselskabet tidskrift Tema: Okker, 5/84, 2-25.

RHEINHEIMER, G. (1991): Mikrobiologie in Gewässer. 5. Aufl., Verlag Gustav Fischer, Jena – Stuttgart.

RIBE AMT & SJA (1997): Laksefiskene og fiskeriet I vadehavsområdet – Resumerapport [Salmons and fishing in tidal areas – Results report].

RINGKØBING AMT (2000): Optimering af okkerrenseeffekten I vinterperioden [Optimization of ochre cleaning effects during winter periods] – Et project udfort af DHI – Institut for Vand of Miljø [Danish Hydrological Institute], Ringkjøbing Amt.

RINGKØBING COUNTY (1995): Ochre removal plant at Spabæk lignite mine, information leaflet.
RÖCKMANN, C. (2001): Von Pyrit bis Schwefelsäure. Universität Bayreuth, Lehrstuhl Hydrologie, Limnologische Station < http://www.uni-hamburg.de/Wiss/FB/15/Sustainability/roeckmann-Dateien/Braunkohle_Forum2001.pdf > (25.07.2005)
SCHEFFER, F., SCHACHTSCHABEL, P., BLUME, H. P., BRÜMMER, G., HARTGE, K.H., SCHWERTMANN, U., AUERSWALD, K., BEYER, L., FISCHER, W.R., KÖGEL-KNABNER, I., RENGER, M., STREBEL, O. (1998): Lehrbuch der Bodenkunde. Ferdinand Enke Verlag, Stuttgart, 494 S.
SCHMITKE, K. D. (1985): Auf den Spuren der Eiszeit – Die glaziale Landschaftsgeschichte Schleswig-Holsteins in Bild, Zeichn. u. Kt.-Skizze.Druck- und Verlagsgesellschaft Husum, 101 S.
SCHÖNBORN, W. (2003): Lehrbuch der Limnologie. Schweitzerbart'sche Verlagsbuchhandlung (Nägele u. Obermiller), Stuttgart, 588 S.
SCHRECKENBACH, K. (2005): Einfluss von Umwelt und Ernährung bei der Aufzucht und beim Besatz von Fischen. Institut für Binnenfischerei e.V., Potsdam-Sacrow < http://www.lfv-swh.de/schrecken1.htm> (12.09.05)
SJA (1993): Project proposal for the installation of the big Hvirlå Ochre Lake, River Section Sønderjyllandsamt, Tønder.
SJA (1995): Project proposal for the installation of Løgumkloster Bæk Ochre Lake, River Section Sønderjyllandsamt, Tønder.
SJA (1996): River Revitalisation – on river restoration in southern jutland. Information leaflet, 09.1996.
SJA (1999): Project proposal for the installation of Landeby Bæk (Kisbæk) Ochre Lake, River Section Sønderjyllandsamt, Tønder.
SJA a (2002): Okker – udvikling i Fe++ (pp-presentation), River Section Sønderjyllandsamt, Tønder.
SJA a (2005): Okker-status, report at Sønderjyllandsamt. Vandløbsafdelingen [River Section], Tønder. Translated by TENT, L. [sinngemäße Teilübersetzung der Entwurfs-Kopie, LT 02.03.2005]
SJA b (2002): Forslag til eTablering af Ravsted vinterokkersø [Project proposal for the instalment of an winter ochre lake at Ravsted]. Projektforslag fremlagt I offentlig høring I perioden 9. Maj til 6. Juni 2002. J.Nr. 8-75-12-6-0, Sønderjyllandsamt, Tønder.
SJA b (2005): Forslag Aktivitetsplan for reTablering af naturlig Laksbestand Juni 2005 i Ribeåsystemet [Suggestion for the Action Plan for re-establishing natural salmon populations June 2005 in the Ribe stream system], Vandløbsafdelingen [River Section] Sønderjyllandsamt, Tønder.
SJA TEKNISK FORVALTNING (1991): Restaurering/naturopretning i Rodådalen [Restoration within Rodå Valley]. J.Nr. 8-08-02-30-89 II, Sønderjyllandsamt, Tønder.
SYMADER, W. (2004): Was passiert wenn der Regen fällt? – Eine Einführung in die Hydrologie. Verlag Eugen Ulmer, Stuttgart, 247 S.
TENT (2001): Pflanzen und ihre Bedeutung für Fließgewässer – Praxistipps – Übersetzt aus dem Dänischen. Edmund Siemers-Stiftung, Hanseatische Natur- und Umweltinitiative e.V., Hamburg.
THORDAL CHRISTENSEN, B. (2004): Plantebestanden I okkerrensningsanlæggene I Herning Kommune [Investigation of vegetation diversity and cover within ochre abatement plants in Herning Municipality].
UHLMANN, D. & HORN, W. (2001): Hydrobiologie der Binnengewässer – Ein Grundriss für Ingenieure und Wissenschaftler. Verlag Eugen Ulmer, Stuttgart, 528 S.

[UBA] Umweltbundesamt (2004): Water Framework Directive – Summary of River Basin District Analysis 2004 in Germany, Umweltbundesamt [Federal Ministry of the Environment, Nature Conservation and Nuclear Safety]
< http://www.umweltbundesamt.org/fpdf-l/2919.pdf > (06.09.2005)

WANDALL, K., WHITELAW CHRISTENSEN, T. (1999): Rapport om okkersoerne i Hvirlå [Report about the Hvirlå Ochre Lake]. Sønderjyllands Amt, Tønder.